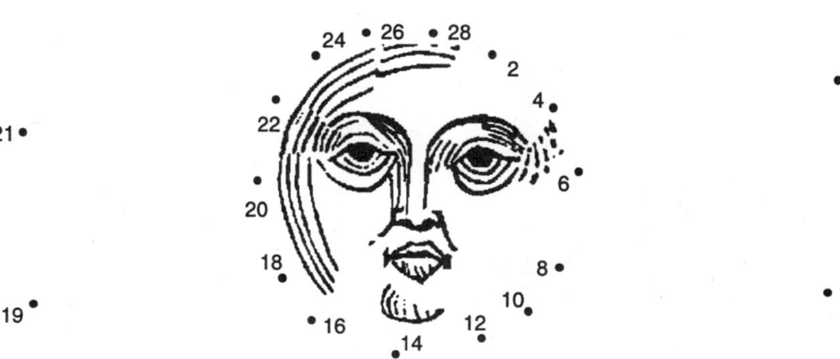

The Secret of Flow
A QI Story

Having initiated focused improvement efforts throughout the land, the King and Queen were ready learn how to sustain the improvements that had begun to appear in even the remotest regions of the land. Something they had seen in their other travels had brought a question to mind. Everywhere else they went, they saw crops standing in fields, granaries filled with the harvest, mills standing idle or working endless hours, carts and ox-drawn wagons waiting idly for crops or flour, and ships lying at anchor in the harbors, some sailing with other goods, but in Robin's lands they had seen no such thing. It was deceptive. It seemed like the crops became grain that became flour that was transported in a seemingly endless flow. Their curiosity got the best of them and with the kingdom's wise man they arrived at Robin's castle on a cool September evening.

Overhead a flock of geese honked as Robin, dressed in Sherwood Green, came out to meet them. "Your majesties," he proclaimed, bowing in respect, "what brings you here?"

The Queen allowed Robin to kiss her gloved hand before responding. "We became curious about the way your harvest seems to flow, where in other parts of the kingdom it seems to rot on the stalk or lie endlessly in storage."

"Yes," the King said, "we suspect you know even more that you haven't told us about the secrets of your crop production system."

"Ah yes!" Robin said, cocking his head to one side and looking up at the first stars of twilight. "I have forgotten something haven't I? But it will take some time to explain it. Do come inside."

Robin led them to a small room off the main kitchen where the warmth of the fires filled the room with cozy feeling, plenty of light, and a soft aroma of wood smoke. They sat in chairs of stout oak and polished leather. Servants brought fresh scones and tea on broad silver platters.

1

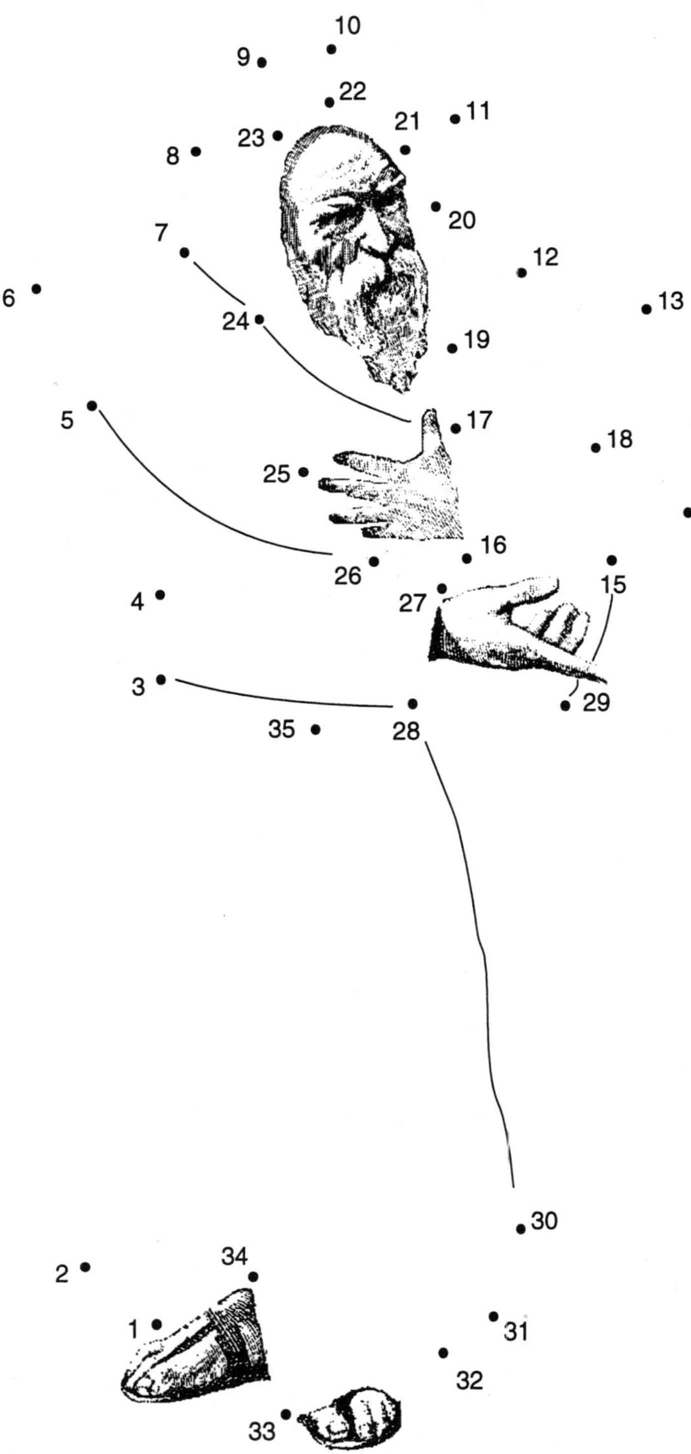

"I so love this room," said the Queen.

"So what is this secret of which we speak?" asked the wise man.

"It is as the Queen surmised," Robin said, seeing the Queen raise an eyebrow at his words. "the secret is one of flow." Seeing the King's brow furrow, he continued. "Imagine a river, a stream, a creek, large or small it doesn't matter, but water flowing through a channel of some kind. How does the water flow?"

"From the highest point to the lowest point," said the King.

"Precisely," Robin replied. "One way to look at it is that the rain falls and pushes the water down hill. Another way is to look at it is as if the water is pulled by the need or demand of the ocean. It is as if each portion of the stream pulls the water from above. And the headwaters pull the rain from the sky."

The Queen shook her head as if trying to shake off a bug. Robin could tell that he'd spoken too much too quickly.

"What do you mean," asked the wise man, "that the river pulls water from above?"

"Let me demonstrate in a different way," said Robin. He grabbed a strip of leather from a nearby table and handed one end to the Queen and pulled it taut. "If I push the strap toward you what happens?"

"It droops," she replied.

"Excellent," he said. "When other fiefdoms try to push their harvest through their system, it creates slack in the system—what we call 'work-in-progress' or WIP for short. Crops pile up, transportation and processing are either overwhelmed or under used.

"What happens now when I pull on the strap?" So saying, he began to pull the strap until it was tight and then just a little more, moving the Queen's arm and hand forward.

"It pulls the strap tight," said the King.

"Exactly! When our people harvest, we use a 'pull' system to keep slack out of the process and WIP out of the system. Ships pull the wagons of flour that pull the mills that pull the carts of grain that pull the harvest." Robin noticed slight unconscious nodding of their heads as the idea began to sink in. "Like a river, the ocean pulls from the mouth of the river and so on upstream. Throughout our harvest, we seek to achieve a flow that is neither drought or flood. Too little is costly and inefficient. Too much is also wasteful. When floods occur in the kingdom, what happens?"

"The water washes away useful topsoil and piles up debris that must be cleared. It leaves water standing in the fields as it recedes," said the King.

The QI Connect-the-Dots Book

"Yes!," exclaimed the Queen, her face blooming in the first moments of understanding. "A flood of crops or goods creates waste."

"And a drought?" asked the wise man, nodding with understanding.

"Parches the fields and limits the yield," replied the Queen who glowed all the more.

"So you see," said Robin, "the way we process the harvest requires cooperation and understanding all of the way through the system. The end must pull the middle and the middle must pull the beginning."

"And the beginning, " said the wise man, "in its own way, must pull the end?"

"Of course," replied Robin. "That way, we avoid the floods and droughts that other's experience during harvest. On the morrow I'll show you how it works."

Robin rose and bowed to the King an Queen. "It will be my pleasure."

A bandy rooster with shimmering red and black feathers crowed the break of day. The sound of pots clanking in the kitchen added to the din and soon the small party were gathered together again.

"If it is all right with your majesties, I suggest we take your coach as it would be more convenient for us all to ride together."

Outside, their sleek, black lacquered carriage pulled by four white horses awaited. After a quick breakfast of porridge and thick dark bread with small portions of pheasant, they climbed into the carriage for a short trip to the harbor nearby.

"This is where it all begins," said Robin. "Aside from the grain we keep for ourselves, we ship fine flour and grain to other parts of England and the mainland. Here at the mouth of the river lie hungry ships waiting to be fed. We seek never to produce one sack more than they require."

Sea gulls hung suspended in the air, their sharp cries filling the thick ocean air. From Robin's vantage point on a small hill overlooking the harbor they could see the wagons unloading sacks of flour and grain into the coastal and ocean going ships.

"I see no warehouses on the docks," said the King.

"There is only a small covered area to hold the work-in-process," replied Robin. "It wastes time to transfer sacks from the wagons to the warehouse, from the warehouse back to a wagon, and then to a ship. By bringing the right number of wagons at the right time, we simply transfer the sacks from boat to ship. The wagons are free to return for more goods. We keep very few wagons here at the docks."

"Amazing," said the Queen, clearly impressed by the efficiency of it all.

"Yes," said Robin, "but not simple to do. I've arranged a code with the incoming ships. Can you see that small schooner coming in from the northwest?" All heads turned to the twin-masted vessel angling through the shallow waves toward the harbor. "We read the flags flying on her mast which show what they want as they sail into the harbor. The wagoneers see the ship's orders and use fast riders or arrows to shoot their orders to the mills. There they immediately begin producing exactly what's needed. The mills order grain and begin processing what is already available. People in the fields begin to harvest the ripened crop."

"Truly amazing," said the King.

"It all begins with a request from the sea which begins to pull the necessary resources from all over the region. Let's follow this ship's request as it moves through the system." Robin herded them back into the carriage and with a few instructions to the driver settled back into his finely textured seat.

As they drove away from the coast, they passed wagons filled with sacks going toward the harbor and twice, riders on horseback passed them going inland. Robin pointed them out as they passed. They arrived at a mill just as the third rider approached. "You've got to see this, " Robin said with a grin.

The rider dressed in simple woolen clothing and carrying a small leather pouch leaped to the ground and ran into the mill. Robin beckoned the King and Queen to follow. The wise man in his flowing, gossamer robes followed.

The rider extracted a simple piece of parchment from the pouch and gave it to the miller, who looked at it for a few moments and then jotted down another order and handed it to the rider. In moments the small man was on his horse; the sound of hooves pounding receded further into the countryside.

"What was that all about?" asked the King.

"The miller took the ship's order and placed another order with a farmer for delivery of the grain to be ground into flour. Soon the farmer will begin harvesting the sacks necessary to load a wagon to the mill."

"Won't they have to wait too long?" asked the Queen.

"No, there's enough sitting here for the miller to start, and it will be replaced by the farmer's harvest when it arrives. Then the mill will process just what's needed and a wagon or two will arrive to load the processed flour to take to the waiting ship."

"It seems complicated," said the King, pushing his crown back to scratch is forehead.

"At first, it seems that way, but then it becomes second nature. Let's continue our journey up the harvest river."

The next stop was the farmer's fields. Here a wagon was already loading grain. Men stood in the fields scything grain while women in simple garments threshed the crop. Boys and girls scooped the grain into bushel baskets for the wagoneers to load.

"This," said the Queen, "looks more like poetry, not work. Do they all help each other?"

"They know it's for the good of all, so they do."

They watched in silence as the wagon filled with ripened grain and then followed it slowly back to the mill. There they found that the miller had just finished the supplies on hand. When the wagon arrived, his helpers began unloading the grain directly onto the grindstone. The wagoneers unloaded the replacement baskets under a sheltered place and began to load the processed sacks of flour onto the scarred oak planks of the wagon's bed. While they waited, Robin and his guests ate a small lunch of cold turkey, fruit, loaves of bread and sharp cheese under a nearby tree. An hour later the wagon began its trek toward the sea.

Less than 45 minutes later, the wagon rolled up to the wharf. The unflappable oxen pulled the cart right out to edge of one pier where the crew began to transfer the sacks into the hold of the ship. A short time later, another wagon joined the first. Just before sunset the ship set it's tops sails as it maneuvered out of the harbor bound for a hungry port.

Robin suggested they have dinner at a nearby inn and stay the night.

They met in the main room for dinner. The table held fine china and silver for the small party. Wine goblets and a decanter of ruby red wine stood on silver tray. A roasted pheasant and a hank of ham steamed on the table. A hogshead of cheese sat waiting to be carved and the smell of freshly baked bread filled the room. Baked potatoes lay wrapped in a towel. They devoured the meal and several pints of ale. In their hunger, no one had spoken for a time. The only sounds were the crackling of the fire in the hearth and the contented sounds of people eating. The wise man used one hand to smooth his beard before he addressed the King and Queen: "Each of you have developed great skills in the course of your lives. I was wondering, what were the most important ones?"

"Leadership and management," said the Queen. The King nodded in agreement and added: "Weaponry and navigation."

"These are core skills, are they not?" asked the wise man. He watched as they glanced first at each other and then back to him. By now, they had learned to take none of his questions for granted. "How did you first learn these skills?"

The King twisted his scepter as he thought. "I studied with the finest teachers."

"And what did they teach you?" the wise man asked softly.

"They taught me how to lead, how to handle the sword and the bow, how to guide a ship to its destination, and how to manage the ebb and flow of the kingdom's finances."

"And I'm curious," said the wise man, "when you say 'how to', what do you mean, specifically?"

The King's face shifted. One eyebrow went up in a bushy arch. He looked for all the world like a young boy who had just been asked the most obvious question in the world, one which he had never thought to ask.

The Queen smiled at the King's expression and tentatively replied: "They taught us, step-by-step how to set direction, manage revenues and expenses, run the castle, and so on."

The wise man rubbed his chin as he nodded in agreement. "Step-by-step you say. It sounds like they taught you the <u>process</u>."

"But there is no process for shooting an arrow," grumbled the King.

"There isn't?" Robin said, a shocked tone in his voice. "For an arrow to find its mark what must you do first?"

"You just pull, aim, and let it fly," rumbled the King.

"Pull, aim, and let fly are three steps of the main process," said the wise man. "But first don't you have to string the bow, select an arrow, nock it on the string, draw the bow, aim, and then release?" The wise man watched as the King's eyes narrowed, but he slowly nodded in agreement. "Did the bow and arrows spring forth whole from the ether or did great craftsmen supply you with the bow and arrows?" The King's eyes narrowed even more, but the light of understanding began to rise in them. The wise man continued. "There is a process for everything that happens; some are just better than others. Some processes we learn so well that we do them without thinking about them at all."

The Queen's eyes brightened and took on a mischievous look. "Even for creating new ideas?"

"From where do new ideas spring?" asked the wise man. "My Queen has created many innovative ways to manage the kingdom. How do you get these ideas?"

"They just come to me," she replied.

The wise man noticed that the King was leaning in, listening closely. "Do you not find that you must first fully understand how things are done now?" The Queen nodded. "And then you explore and learn other areas of knowledge to gather the seeds of the new idea?" The Queen nodded again. The mischievous look began to shift to one of intense curiosity. "Then do you not explore the similarities and differences between the old and the new, looking for ways to do things more easily?" Again she nodded. "And do you not find that most of this comparison and revelation occurs when you are thinking about something else entirely?"

The wise man watched as the King leaned back in thought and the Queen's face softened in understanding. "Just because we can do something easily doesn't mean that there isn't a process behind it."

"Are you saying that we need to understand our processes?" asked the Queen.

"Precisely," replied the wise man, "and measure them so that we can begin to make significant improvements in the time they consume and the waste and rework they generate."

"Measurements!" huffed the King. "It all sounds quite boring."

The wise man smiled. The King had once again blundered into quicksand of common thinking. "To someone unfamiliar with process improvement, understanding processes and measurement might seem too boring, constraining their thinking and possibilities. Are you not, my lord, always interested in improvements in weaponry, music, or the arts. Are you not here to learn more about how to lead more effectively?" The King shrugged and nodded.

"I believe we are all engaged in the quest for mastery," The wise man continued. "Being successful year in and year out depends on the ability to do a task consistently and repeatedly. For many of the best farmers, millers, and crafts people, this ability comes from years of experience. They have become masters in their field, but they don't know how they do it. They just know that they can. To be able to pass on this wisdom more easily, we must begin to capture and understand the process that each master follows to accomplish their tasks. Then we can begin to learn how each task feeds into the next so that we can continuously improve the collective processes of our whole community."

"It seems too simple," the King grumbled.

"Of course," said the wise man, "the best approaches are always simple. Complexity is a trap to be avoided at all costs." Then he fell silent again and signaled the proprietor for another pint of ale.

Robin wiped the remnants of the meal from his chin and called for some parchment and a pen. When they arrived, he began to write and speak at the same time. "I always find it best to display my thinking on paper. Only a fool tries to work everything out in his head. To understand the flow through the Kingdom, I suggest we begin by identifying the core processes."

"When you say 'core processes'," said the Queen, "what do you mean?"

"I found that there are three types of core processes: 1) management and leadership processes which guide all efforts, 2) development and delivery processes for products and services, and 3) support processes that aid the development and delivery of products and services."

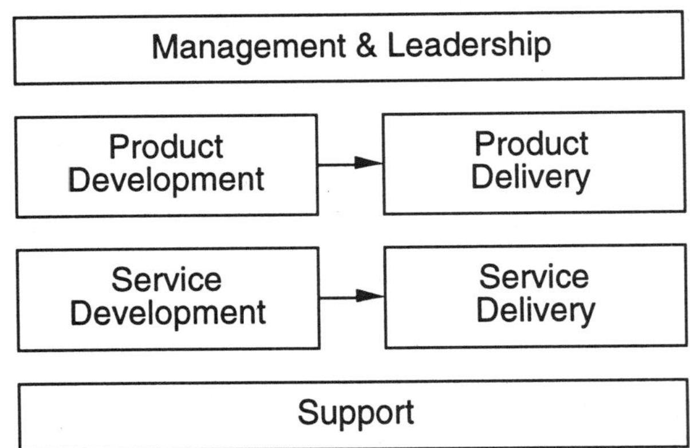

"Now let me ask you, what are the most important issues facing the Kingdom?"

The King and Queen bent their heads together, speaking softly for a few minutes. The wise man watched their unconscious communication of gestures and looks. When they nodded in agreement, he glanced away as if distracted by the room, but from his peripheral vision he noticed the King about to speak.

"We would like to increase our kingdom's ability to produce crops, livestock, and goods. More food would feed our growing population. More goods would allow us to grow our exports and increase the wealth of our nation and our subjects."

"And which of these is most important?" asked Robin.

"Crops," said the King, "as we saw today in the fields." The Queen nodded and said: "Without crops, we could not feed our people, raise as much livestock, or have the raw materials for our crafts and leather goods."

"So we must first focus on the overall crop process—the step-by-step process for planning, planting, growing, harvesting, and processing the crops. And which crop is the most important crop of all."

"Maize," they said simultaneously. "We grind it into flour for our bread and it is our major export," said the King. "We use it to feed our herds and flocks," sang the Queen.

"So let us begin by focusing on the yearly maize process. I have found that virtually all processes involve four common phases: planning, doing, checking the result, and acting to improve. Let's begin by drawing the overall process for maize."

"Shouldn't we start with just the growing process?" demanded the King.

"People often work on each independent aspect of the overall process. Unfortunately, when one step of the process improves dramatically, the next step may stumble. Let me see if I have a good example."

Robin walked to one of the large floral arrangements in the room and adjusted a few flowers and reeds to give it a more harmonious appearance. After removing a few wilted flowers and leaves, he returned to his chair and began to speak. "As you saw today, our overall process involves the delivery of milled flour and maize for livestock. This involves planning how much to plant, when to plant, preparing the soil, planting the seed, tending the crops, harvesting the grain, storing the grain, milling the grain, and delivering the milled flour. If we plant too much, there won't be enough storage space or milling capacity; the grain will go to waste on the stalk. If we increase the storage and milling capacity, but not the harvest or delivery capability, then we have idle resources, which is wasteful. How many times must this happen before we learn how to look at the entire process of creating milled grain. Each individual process is part of a larger process. Looked at collectively, improvements require the logical connection of all of these elements, not just an isolated few."

"And that's why we must focus our efforts on a few 'core' processes?" asked the Queen.

"Yes!" he replied. "We cannot hope to do them all, but we can focus on the most important ones first. Core processes have evolved and grown as a reaction to problems. Where once our core processes were fast and simple, they have become increasingly slow, tedious, and complex. If, for example, a late snow kills a young crop, then farmers begin to wait until the later in the spring to plant anything, thereby delaying or reducing the crop. The more we look, the more we will discover that our processes are filled with laborious activities to deal with waste and rework. We are paying the high price of not doing it right the first time."

"By discovering how these processes work ," Robin continued, "we often find obvious ways to simplify and streamline the processes. Furthermore, we will find simple ways to track how each process behaves so that we will know if it is working properly or needs improvement. Like a river that flows through a series of channels and lakes, work also flows through various processes. We can diagram the flow with boxes to show activities or diamonds to show decisions, and connect the boxes with arrows.

Yearly Maize Process

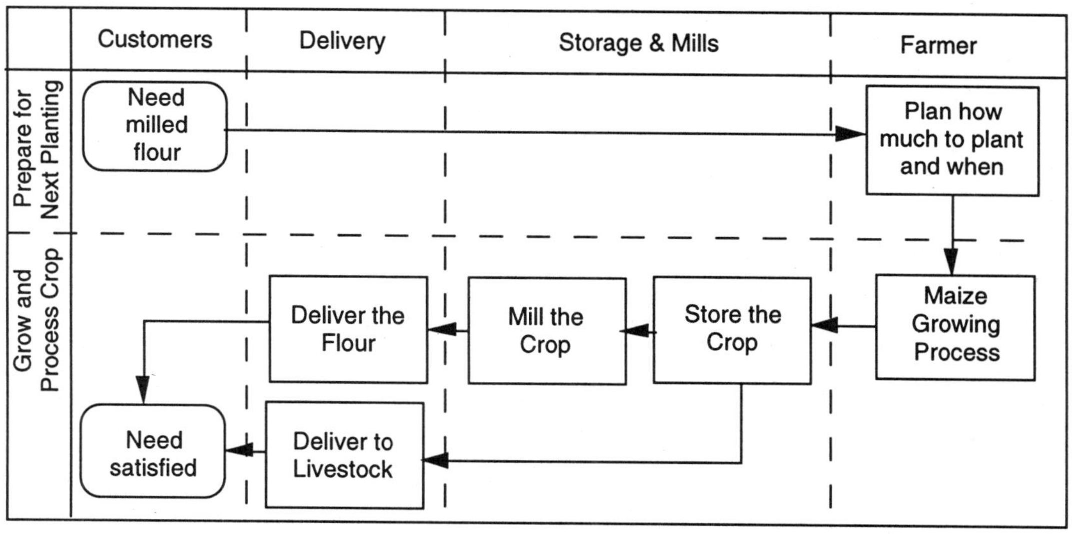

"Then, within each overall process, there are other sub-processes like the maize-growing process:

Maize Growing Process

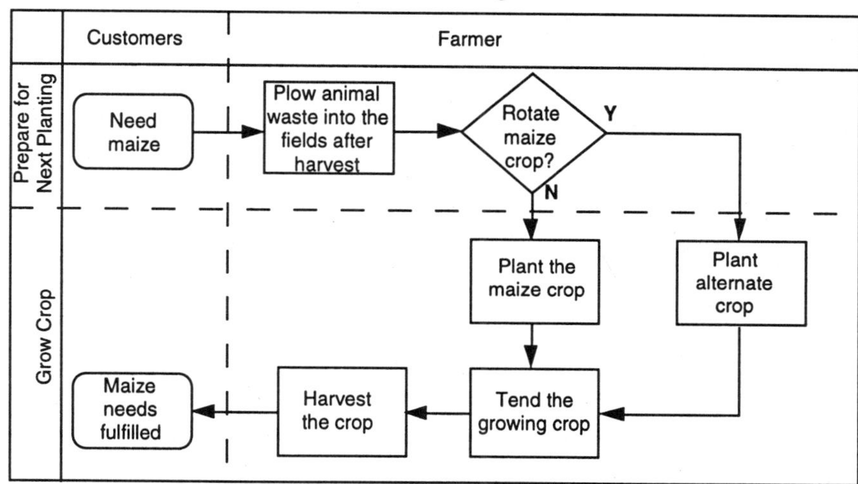

"Next, analyzing the process will often show significant ways to improve. I use a simple checklist to understand how long each activity, decision, or transfer takes, and the added value of each one."

"What do you mean?" asked the Queen. "What is value added?"

"We add value to the product or service whenever 1) the customer cares if we do an activity or step, 2) our actions change the product or service in some useful manner, and 3) it's part of doing it right the first time. As you can see from this checklist, only a few of our maize growing activities are truly "value added."

Value Checklist				
ACTIVITY, DECISION, or ARROW	TIME SPENT (hours, days, weeks, months)	NON-VALUE ADDED?		
		IDLE TIME? WAIT TIME? DELAY?	INSPECT? DETECT? TEST?	REWORK? WASTE? SCRAP?
◯ Need for Milled Flour	0	N/A	N/A	N/A
→	0	Y	Y	Y
▭ Plan how much to plant and when	0-4 months	N	N	N
→	0	Y	Y	Y
▭ Maize Growing Process	3-4 months	N	N	N
→	1-3 weeks	Y	Y	Y
▭ Store the Crop	4-6 weeks	Y	N	Y
→	0	Y	Y	Y
▭ Deliver to Livestock	3-5 days	N	N	N
→	0	Y	Y	Y
▭ Mill the Crop	1-3 days	N	N	N
→	0	Y	Y	Y
▭ Deliver the Flour	3-5 days	N	N	N
→	0	Y	Y	Y
◯ Need satisfied	0	N/A	N/A	N/A

"I didn't see much of it today, but what about storing the crop?" demanded the King. "That adds value doesn't it?"

The wise man smiled and winked at Robin. "Do our customers care if we store the crop before we deliver it?"

The King grumbled and frowned. "No," he said softly, like the sound of distant thunder.

Seeing where Robin was headed, the wise man asked: "Does it physically change the maize?" Both monarchs shook their heads. "Is it part of doing it right the first time?"

"No!" sang the Queen. "It's so obvious, but we've always done it that way. I'm curious about the arrow before. Why does it take up to three weeks to get the maize to storage?"

"In the past, when everyone plants and harvests at the same time, there aren't enough people or carts to bring the harvest in. Ripened maize can sit on the stalk for weeks before being harvested and stored."

"Either way," said the Queen, "the maize sits idle."

"And an idle product wastes time and money," said Robin. "So there are some obvious ways that this process could be improved. You saw some of them today. What are they?"

"What if we staggered the plantings?" asked the King. "Wouldn't that also stagger the harvests so

  The QI Connect-the-Dots Book

that we could minimize the time the maize spends on the stalk?"

"And wouldn't that also make it possible to deliver the maize without storing it?" asked the Queen. "Wouldn't that save time, waste, and the rework of loading the maize twice?"

"Of course," said Robin. "As you can see, just by examining a process in this way, simple, elegant improvements seem to leap off the page." So saying, he began to draw.

Yearly Maize Process

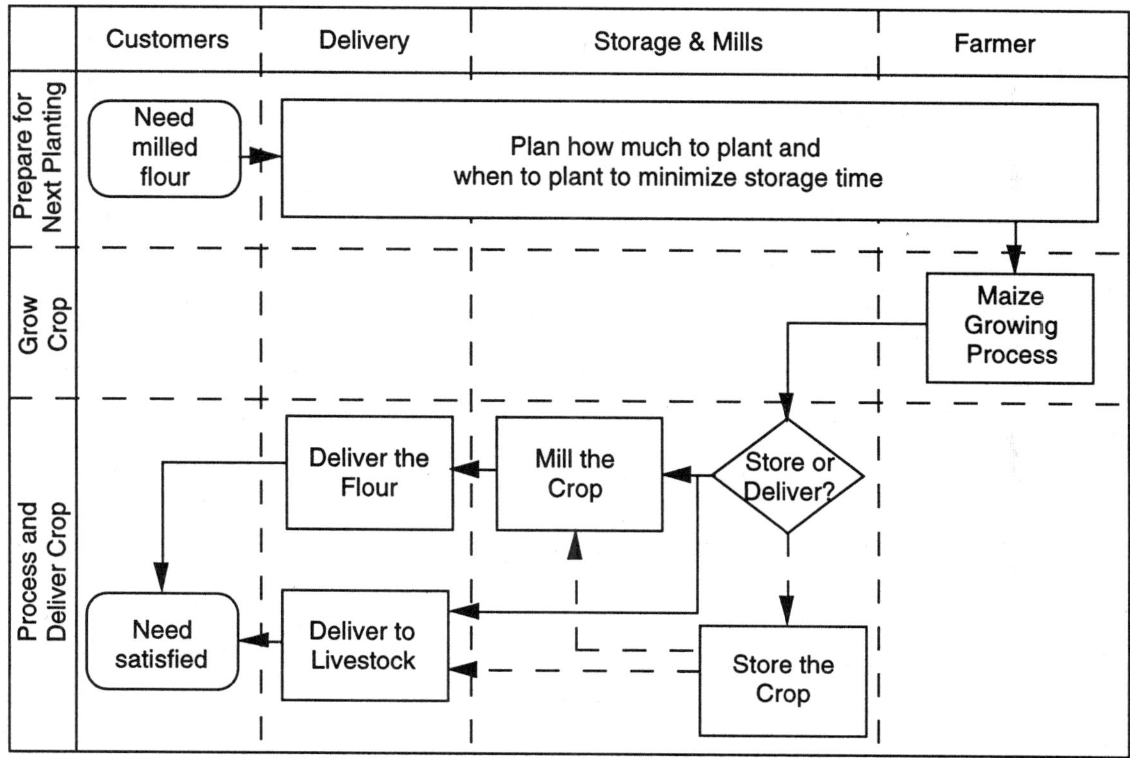

The Queen perked up. "That seems simple enough, but how would we begin to understand and improve these processes?"

"I have discovered that there is always some natural variation in any process," Robin continued. "What I mean is that no matter how consistently we do things, it will always take a little longer or shorter to grow a crop from one year to the next; there will always be some variation in the amount of rain or pests or whatever. To explain variation, let me ask the King: Remember when you first learned to shoot the bow at a target." The wise man watched as the King went back in time. "Describe the target to me."

"It was circular and had three rings of equal width."

"When you first learned to shoot, what did you aim for?"

"I seem to recall being so concerned with notching the arrow and the mechanics of the pull that I don't remember aiming. I just remember how excited I was when I first hit the target."

"Precisely," Robin said. "Were you then able to repeat your success?"

"With time and practice" said the King.

"And when you could hit the target consistently, where did your arrows land?"

"Most would land near the bull's eye," said the King, " and the rest were scattered around the target almost at random. Occasionally I would still miss the target, sending the arrow into the heather behind."

"Precisely," said Robin. "That's natural variation. No matter how consistently you notch the arrow and pull the bow, there are differences in the arrow and the aim and the release and the wind that will change the result, even though most will strike the target. Now let me ask the King: What where you aiming at when you first learned to shoot?"

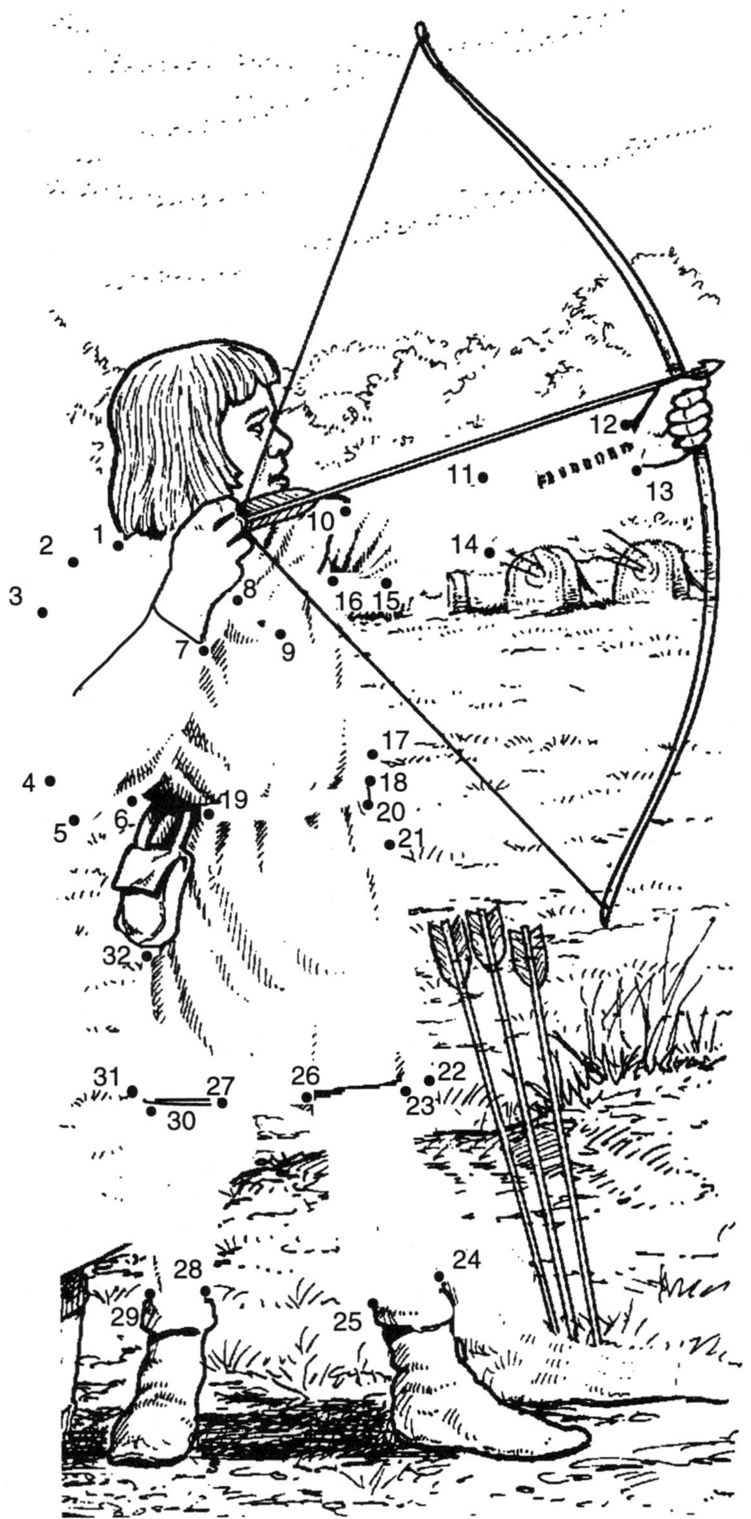

"I aimed at the target. What do you mean?"

"Did you aim at the whole target, or a specific point upon it?"

"Ah," sighed the King. "I see what you mean. I just tried to get my arrow to land anywhere on the target, not anywhere in particular. It was not until I began to aim for the center circle that my results began to improve."

"This is why having a specific target is so important to reducing variation. When we pick specific targets, people will learn how to achieve them and most will come closer to the bull's eye."

"And what have you learned from your study of variation? asked the Queen.

"Simply this," said Robin. "That performance—measured in time, defects, or the cost of waste and rework—varies like arrows striking the three rings of a target. As long as we hit the target consistently, our processes are behaving normally. If we miss the target, we need to investigate why we missed: Was our aim wrong? Were our tools out of alignment? What caused the miss? Similarly, if all of our arrows strike the outer ring or the inner bull's eye, we need to understand what's causing it. Or if our arrows are moving from the center out or the edges in, we need to understand the trend."

"And you have found ways to understand this variation in processes?" asked the King.

"Yes," said Robin. "With experience, it's possible to just look at a simple graph of our performance and see trends and problems. But to be more exacting, a simple set of calculations used on as few as 20 measurements can tell us a great deal about how stable and capable our processes are of meeting the needs of our customers."

"Can you give us an example?" asked the Queen.

"I'd like to show you an example from the fields, but please remember that everyone from shoemakers to innkeepers could use these as ways to measure the daily variation in their processes. It takes, on average, 90 days to grow a maize crop from seeding to harvest. Depending on the weather, rainfall, soil conditions, and so on, it can take as little as 72 days or as long as 108 to bring in a crop. The chart we use looks like this:"

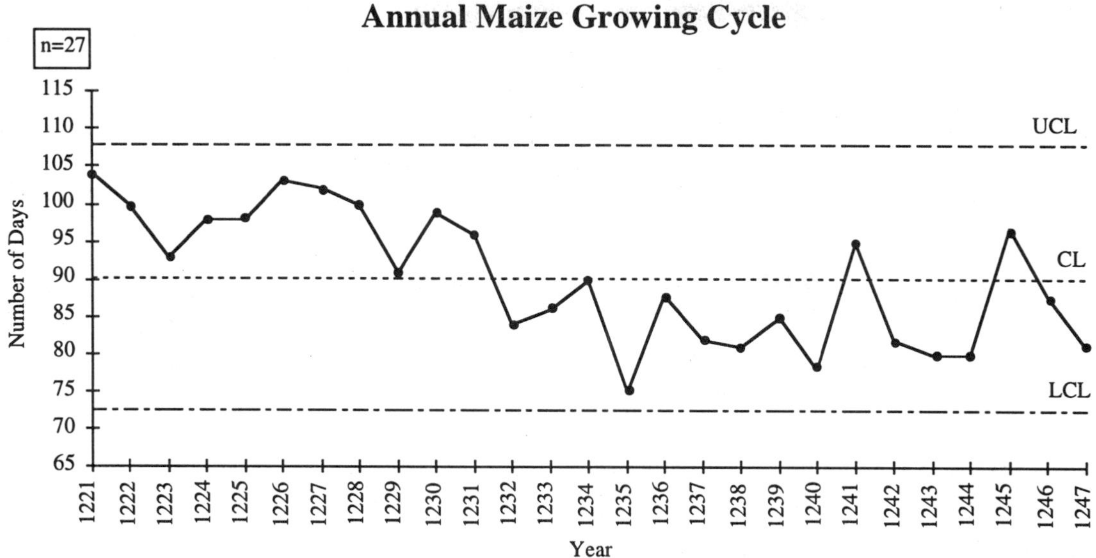

"As you can see, the last 27 harvests have consistently arrived within these limits."

"These are limits that you've set?" asked the Queen.

"A good question, my Queen," said Robin. "But no, these limits are set by the data from the harvests themselves. The process tells me what the limits are. We cannot legislate them. What we can legislate, however, is finding ways to change the process that will improve our results."

"So this process is stable," said the King, "but you also mentioned process capability. How will we know if a process is capable? Can we read capability from this chart?"

"It is possible to read capability from these charts," said The wise man, "but it is much easier to see when using a type of bar chart called a histogram."

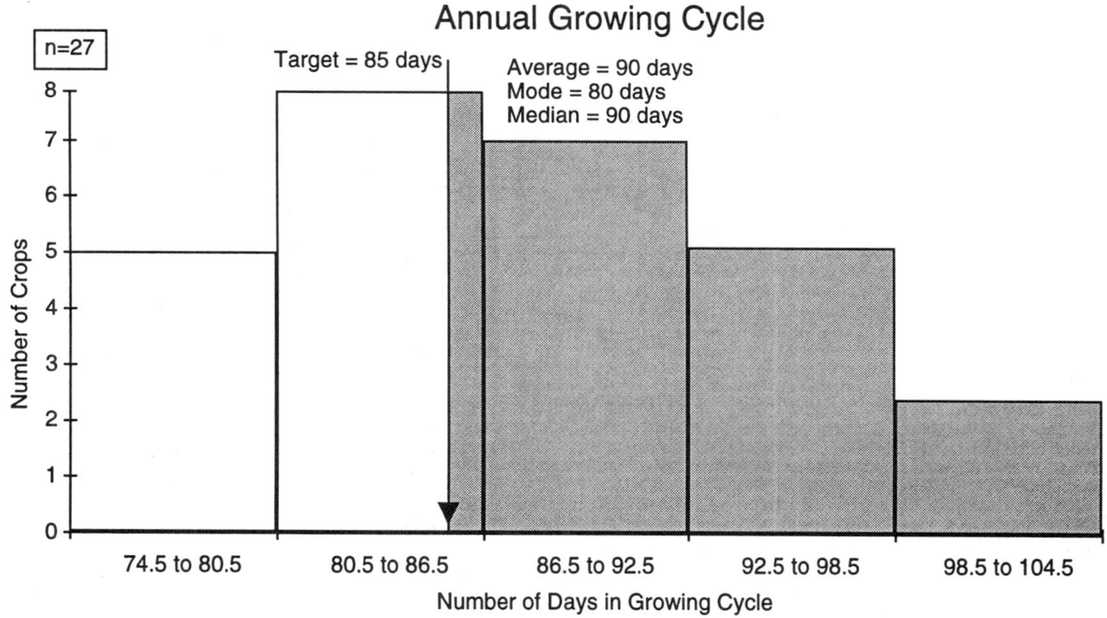

"With this information in hand, it is possible that we could get two harvests a year where we currently only get one. From this chart you can see that if we could consistently grow crops in 85 days or less, that we could achieve a two crop harvest."

"Wouldn't that also require a doubling of storage and milling capacity?" asked the King.

"Perhaps!" exclaimed Robin. "Our mills are idle much of the year. Two harvests a year could keep them busy. If we could deliver two crops a year even some of the time, it would dramatically increase our total ability to produce milled flour. It is possible, however, that we would need more storage."

"Do you think we could sell all of this production?" asked the King.

"With a faster process and the efficiencies of more fully using all of our resources, our maize will be less expensive and of higher quality. I have little doubt that we could find markets for our produce," said Robin.

"This all seems well and good," said the Queen, but couldn't this cause more crop failures because early plantings could die of frost and late ones be destroyed by bad weather?"

"An excellent question my Queen," said Robin. "So we would also need a way to track the percent of the crop that is lost due to weather, would we not? We will want to maximize our harvest while minimizing the waste associated with doing so. We cannot weigh the crop lost as shoots in the spring. We don't always plant the same number of fields each year, so how would we best measure crop loss?"

"Could we count the number of acres planted?" asked the King. "Then would the percent of total acres planted be a good measure of failures?" asked the King.

"Excellent, my lord! Excellent! "

"Is there a way to show variation in this kind of measurement?" asked the Queen.

"Of course," replied Robin as he began to draw.

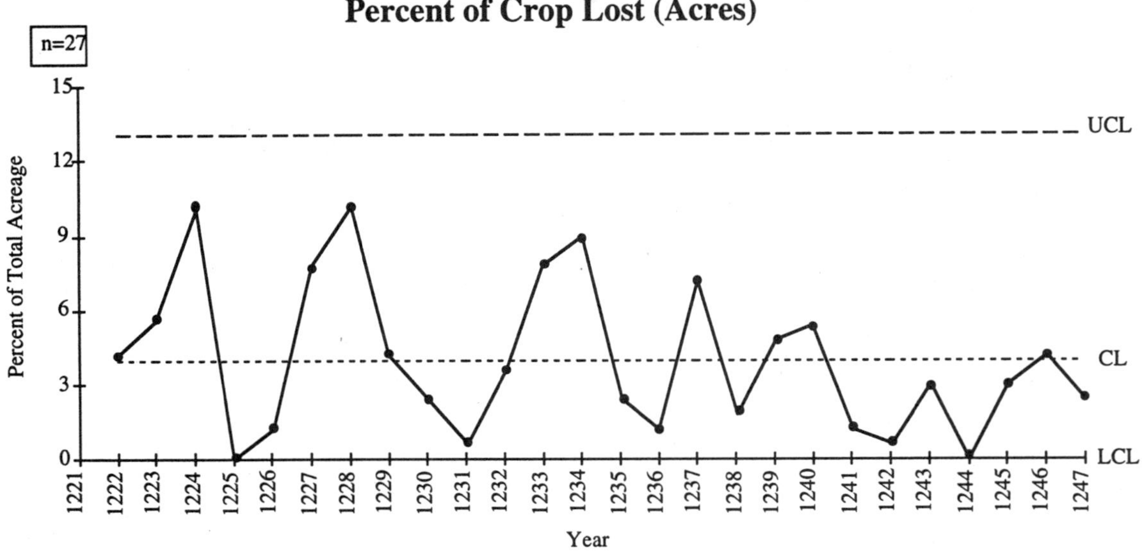

"And what about capability?" asked the King. "Do we use the histogram again?"

"No," said Robin. "If you were the customer, how many crop failures would you want?"

"None," sang the Queen.

"Exactly," said Robin. "A fully capable process delivers zero defects—no waste or rework."

"Impossible!" exclaimed the King.

"Difficult," said the wise man, "but not impossible. Do you not know at least one craftsman who delivers perfect goods each and every time with little or no waste or rework?" The wise man watched as the King and Queen shifted uneasily in their chairs, but finally nodded in agreement.

"Given enough time," Robin continued, "we can achieve extremely low failure rates."

"But, what do we do when the process isn't stable or capable?" asked the King.

"We must seek and eliminate the root cause of the unstable condition. Once the process is stable, we can begin to improve its ability to meet the customer's requirements."

"When you speak of the root cause," growled the King, "what do you mean?"

"Let me ask you: When there is a weed in the garden, what do you do?"

"Pull it up, of course," said the King.

"What happens if you only succeed in snapping off the top?"

"The weed returns."

"Of course," exclaimed the Queen, "whether a weed in the garden or a problem in the kingdom, we must get down to the root of the problem."

"Precisely my Queen," said The wise man. "Far too often, people only succeed in treating the symptom—the visible portion of the problem. This is much like cutting off the top of the weed. This leads to rework because we must periodically trim the weeds. Then we build this weed cutting activity into our process and, over time, our processes get slower and slower burdened by these additional steps. This knee-jerk reaction to the symptom deludes everyone into the belief that the problem is solved. To get to the root of the problem requires more time and careful analysis, but it can prevent the problem forever."

"I understand how to find the root of a weed," said the King, "but how do we find the root cause of a problem in our process?"

"The simplest way is to begin to ask the question: 'Why?' We take the answer to that question and then we ask 'Why?' again up to five times. This usually leads us to the root cause. Now, I know that we had significant problems this spring. What percentage of total acres planted were lost?"

"Well over fifteen percent," said the Queen, sadly.

"So," said Robin, "fifteen percent is well beyond our upper control limit for lost acreage. So let us begin: Why did we lose so much acreage?"

"A late frost," growled the King.

"And why did a late frost kill so many acres?" asked Robin.

"It just did!" snarled the King.

The Queen appeared thoughtful. "For the preceding years we had been lulled into a false sense of security. We were planting more and more acres earlier in the year. Is the answer that we planted too many acres too early?"

"Let's try it," said Robin. "So we planted more earlier because we were lulled into a sense of security from prior years. Why were we lulled into security?"

"Because we forgot that weather varies?" asked the Queen.

"And what caused us to forget?" asked the wise man.

"It's not written down anywhere?" suggested the King.

"Precisely!" exclaimed Robin. "We don't have a process to anticipate weather problems and prevent loss of crops. We can't change the weather, but we can change our process." With that, he took a fine goose quill and began to draw.

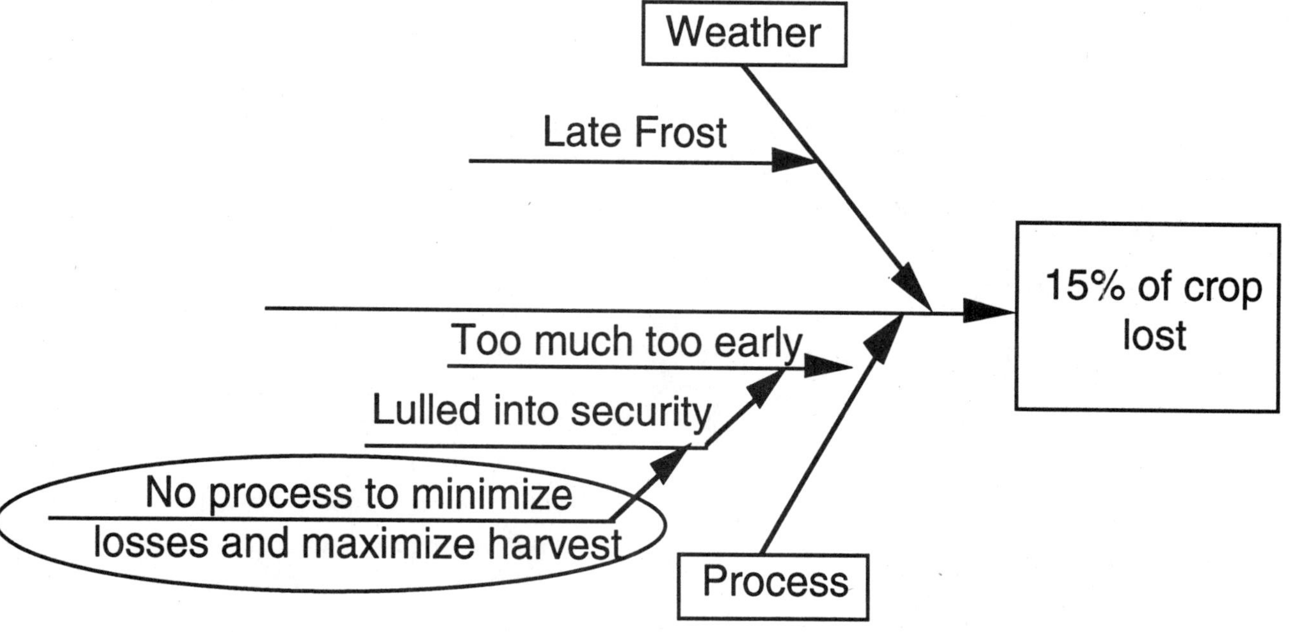

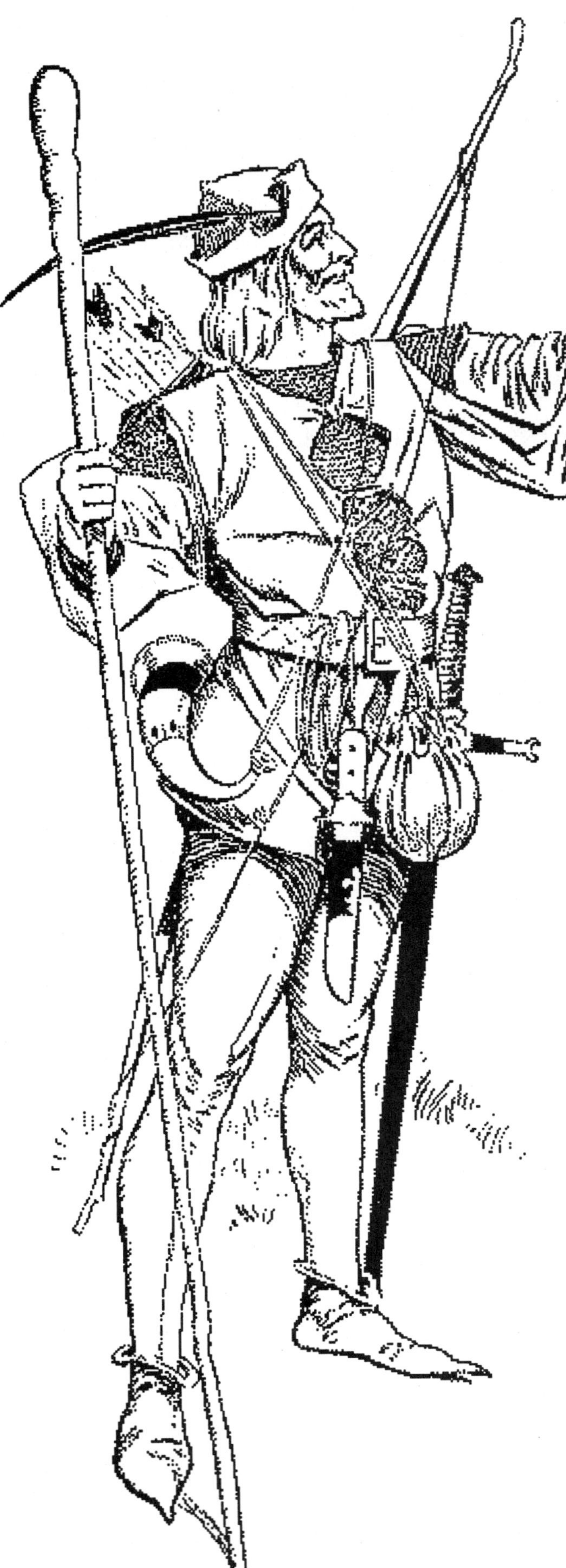

"Now," he continued, "given our earlier discussions, what might we do to prevent this problem in the future?"

"Stagger our planting!" sang the Queen. "Yes," said the King more cheerfully. "Then we can minimize our risk, distribute our load at harvest time, and even squeeze in an extra crop each year."

"Amazing, isn't it?" asked the wise man. "By simply analyzing and changing our process we can begin to reduce the waste and rework while simultaneously increasing the gain. By studying these abnormal conditions we can make changes that will return the process to stability."

The Queen smiled, and then her eyebrows began to dip. "But what do we do when the process isn't capable of meeting our needs? What do we do then?"

"If the process is capable of meeting our needs some of the time, as it is with the maize growing process, then we can commission problem solving teams to look into the common causes of slower growth. One team could look at total cycle time to grow a crop; another could address the common causes of failure."

"I have found that continuous improvement requires a balance of incremental, evolutionary improvements and dramatic, revolutionary improvements. Based on process capability, some of our teams should focus on incremental process improvements while others focus on achieving revolutionary ones."

"So," said Robin, "we must begin focusing our energies and start teams to define and measure the core processes of the Kingdom. Initially only a few key participants will be engaged in defining and measuring the core processes. Rather than engage the whole Kingdom, we can start with one village and then expand."

"Then," said the King. "we must start a variety of teams each year to manage and improve processes throughout the kingdom?"

"Yes," Robin replied, "everyone, ultimately, will be involved in maintaining the gains from prior years by applying the processes and monitoring the indicators."

"Then we can initiate additional problem solving teams?" asked the King as he moved around the table.

"Yes," replied Robin gesturing for the King to be seated, "as we work on the processes, we often find problems to be solved and waste to be eliminated. Once the processes are clearly defined and measured, we can begin to determine where and how to improve them.

The Queen's face wore the "ah ha" look of sudden understanding. The King was combing his beard thoughtfully. "So," Robin continued, "using the process measurements, we will identify specific areas for improvement that will make the greatest contribution to the Kingdom. Problem-solving teams can identify the root causes of the problem, and make the improvements necessary."

"Here, here!" said the King, raising his goblet. They all drank to the success of their efforts. Servants brought in sparkling decanters of brandy.

"Once we have defined, simplified, measured, and tested our core processes," Robin continued, "they can be transferred immediately to the rest of the kingdom because we will have clear, concise processes. This will require careful support from your majesties, because the other lords and ladies will have to be educated to understand the origin of these improved processes. Otherwise it will seem like just another knee-jerk reaction.

"Then, we can work together to develop a plan for improvement. It may take a few years, but with time, we can weave this system of continuous improvement into the very fabric of the kingdom."

"This is a legacy you have chosen to leave for all of the people, is it not?" asked the wise man. He watched as the King and Queen both nodded in agreement.

The fire had died down. Silence filled the inn. "It's late," said Robin. "Perhaps you would like to sleep on what you have learned and as you sleep it will become even clearer and more understandable. Tomorrow morning, after a good breakfast, we will begin initiating the efforts necessary to create a bright and exciting future for our customers, the Kingdom, and our people!"

Managing Processes
Sustaining Improvement Made Easy

Process Management

There is always a best way of doing everything.
  -Emerson

Before we can find better, faster, and cheaper ways of serving customers, we first have to define and stabilize the current way of doing business. Making existing processes predictable and capable of meeting customer requirements follows the FISH process--Focus, Improve, Sustain, and Honor.

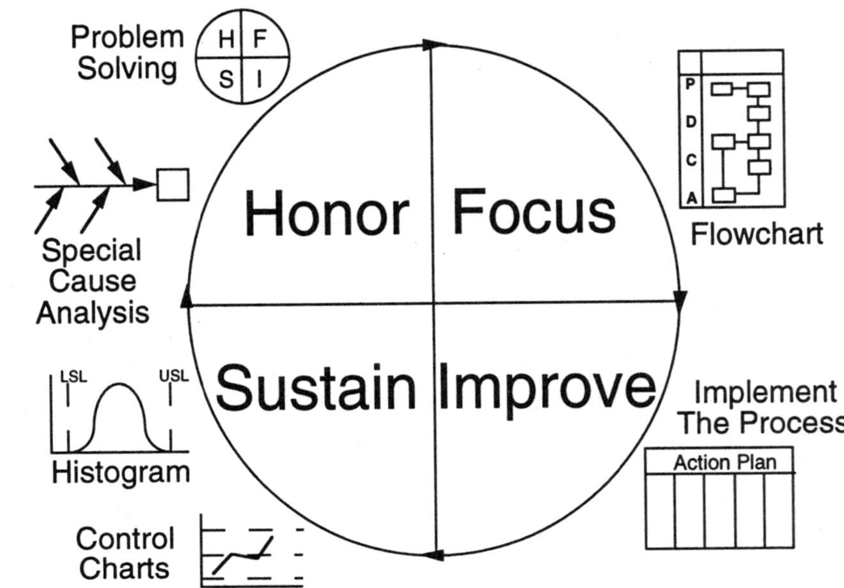

Process

FISH	Step	Activity
Focus	1	Define the process
	2	Identify the "quality" and "process" indicators (pp. 30-33)
Improve	3	Implement the process and indicators
Sustain	4	Check the process for stability and capability
Honor	5	Recognize, review & refocus

Managing Processes
Sustaining Improvement Made Easy

Why Quality?

All progress is based upon a universal innate desire on the part of every organism to live beyond its income.
  -Samuel Butler

Customers have uncanny instincts. If they don't get what they want, most will quietly abandon your business and go elsewhere. Only one customer in eight will complain if they have a problem--every complaint is an opportunity to create customer loyalty. A dissatisfied customer will tell 16 other people; a satisfied customer will tell less than eight. It costs five times more to get a new customer than to keep an existing one. Only *satisfied* customers can create jobs, profits, and successful businesses. How do we deliver what customers want and need?

Customer Requirements

Customers have three common requirements for products and services. To improve delivered quality, each requirement must be measurable. A product or service must be:

Requirement	Specific Examples
1. "Good"	• right the first time (accurate, reliable) • easy to use or understand • treat me like a valued customer
2. "Fast"	• give me what I want when I want it • meet your commitments • fix it fast if it breaks
3. "Cheap"	• charge fair prices • help me save money and make more • eliminate waste and rework

Any man can make mistakes, but only an idiot persists in his error.
  -Cicero

In a personal or business relationship, when you fail to meet your customer's ever-increasing requirements for timely, low-cost, defect-free delivery of a product or service, you have a problem. It will not vanish through magic or good luck. You can spend a lot of time *reacting* to the problem, or you can invest in preventing it. The first step toward preventing problems and retaining the customer's faith in your product or service is to <u>define the process</u>. Then we can remove the non-value added steps, identify the measurements required to manage and improve the process, and implement it.

Purpose

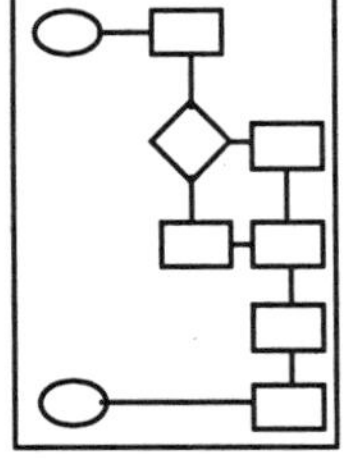

Define the <u>existing</u> process as a starting point to begin improvement

To begin to manage and improve a process, we need to develop a flow chart of the <u>existing</u> process. Unlike simple flowcharts, a process flowchart shows the major steps of the process along the left-hand side. Since processes often follow the PDCA cycle, this flowchart uses it along the left-hand side. Across the top, the flowchart can be divided to represent the individuals, groups, or departments that perform portions of the process. A process flowchart always begins and ends with the *customer* and their requirements.

Flowchart Symbols

The unified process of drawing and shooting was divided into sections: grasping the bow, nocking the arrow, raising the bow, drawing and remaining at the point of highest tension, loosing the shot.
  - Eugen Herrigel

A flowchart uses a few simple symbols to show the flow of a process. The symbols are:

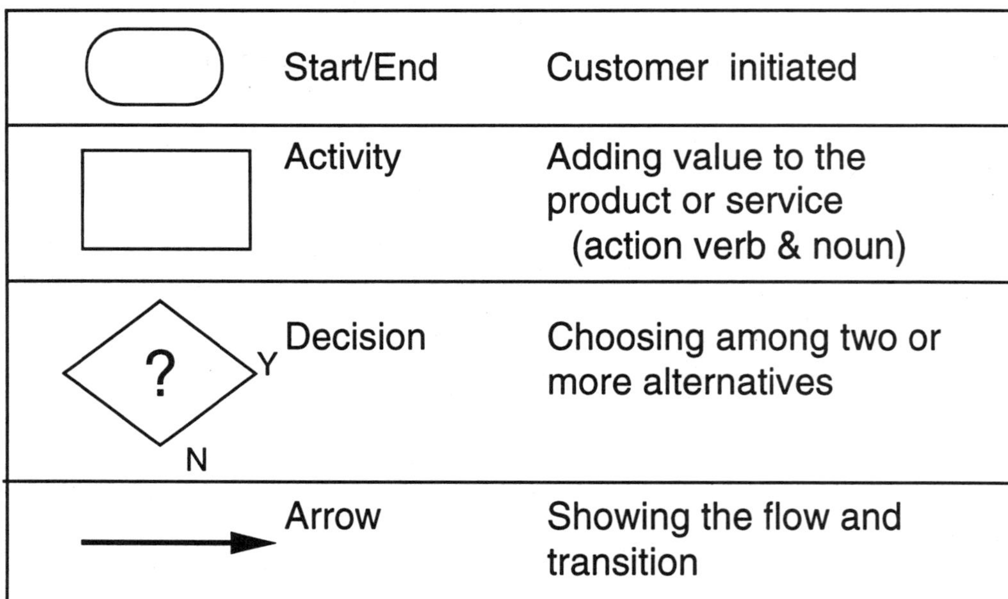

Instead of writing directly on the flowchart, use small Post-it™ notes for both the decisions and activities. This way, the process will remain easy to change until you have it clearly and completely defined. Limit the number of decisions and activities per page. Move detailed subprocesses onto additional pages.

Who / Step	Customer	FLOWCHART		
Plan	Request for product or service			
Do				
Check				
Act	Delivery of product or service			

The QI Connect-the-Dots Book

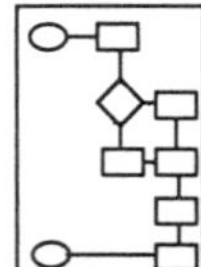

Process Type

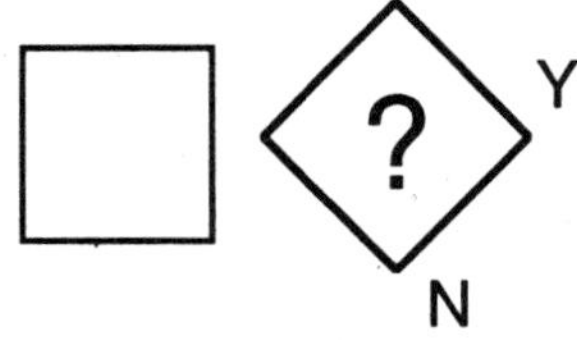

There are four main types of processes: design and delivery of a product or service, management of design/delivery, and support. Trying to show all of these on one flowchart leads to confusion. Diagram each one independently and then show how they're linked. Delivery processes may include everything from ordering to repair.

Brainstorm The Flow

One of the simplest ways to define a process is to define the process is to brainstorm the activities and decisions, and then sequence them on the flowchart:

1. Using <u>square</u> Post-it™ notes (use squares as <u>activities</u>, rotated into a diamond, they represent a <u>decision</u>), begin listing activities (verb-noun) and decisions.

2. Once you have all of the notes on a sheet of easel paper, order them, from the first to the last. Use arrows to connect the activities and decisions.

3. Then identify the group or department that does each set of steps on the top of the flowchart.

4. Identify the major steps of the process on the left side.

Become The Product Or Service

Another way to define the process is to <u>become</u> the product or service flowing through the process. Whether your customer wants a document, a widget, a clean house, a hot meal, a check cashed, or whatever, become the thing going through the process. Then, step by step, go through the whole process until the customer is satisfied.

Where do you start? Who do you go to next? What do they do with you at each step along the way? How long do you wait at each step for someone to process you? At what points do you have a choice of paths to follow? What is the decision you make at these points?

Step 1 - Define the Process
Flowchart the Process

Who / Step	Customer	FLOWCHART		
Plan	Request for product or service			
Do				
Check				
Act	Delivery of product or service			

Step 1 - Define the Process
Value Analysis

Purpose

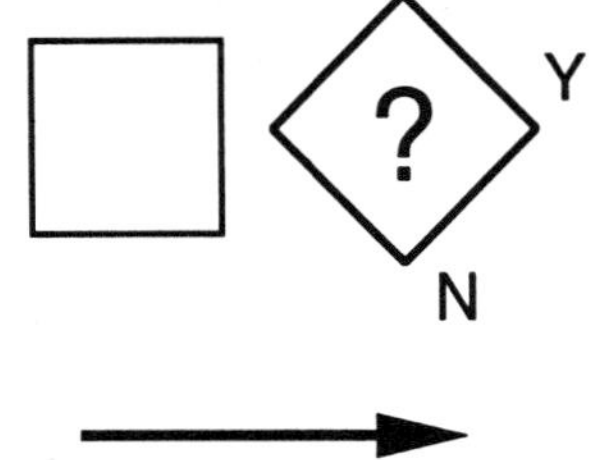

Identify the non-value added activities that can be eliminated from the process.

Over time, processes become cumbersome, inefficient, and ineffective. This complexity consumes more time and accomplishes less. Each activity, decision, and arrow on the flowchart represents time and effort. From the customer's point of view, little of this time and effort adds value, <u>most of it is non-value added</u>. From their point of view, delay and rework do not add value. We can increase productivity and quality by simplifying the overall process--eliminating delay and the need for rework.

Value Analysis

What we must decide is perhaps how we are valuable rather than how valuable we are.
-Edgar Z. Friedenberg

Everything is worth what its purchaser will pay for it.
- Publilius Syrus

What may be false in the science of facts may be true in the science of values.
- George Santayana

Step	Activity
1.	For each arrow, box, and diamond, list its function and the time spent (in minutes, hours, days) on the checklist.
2.	Now become the customer. Step into their shoes. As the customer, ask the following questions: • Is the order idle or delayed? • Is this inspection, testing, or checking necessary? • Does it change the product or service in a valuable way, or is this just "fix it," error correction work or waste?
3.	If the answer to any of these questions is "yes", then the step may be non-value added. If so, can we remove it from the process? Much of the "idle," non-value adding time in a process lies in the arrows: Orders sit in in-boxes or computers waiting to be processed, calls wait in queue for a representative to answer. How can we eliminate delay?
4.	How can activities and delay be eliminated, simplified, combined, or reorganized to provide a faster, higher quality flow through the process? Investigate hand-off points: how can you eliminate delays and prevent lost, changed, or misinterpreted information or work products at these points? If there are simple, elegant, or obvious ways to improve the process now, revise the flowchart to reflect those changes.

Value Checklist				
☐ ◇ → **ACTIVITY, DECISION, or ARROW**	**TIME SPENT** (hours, days, weeks, months)	**NON-VALUE ADDED? (Y/N)**		
		IDLE TIME, WAIT TIME, OR DELAY?	**INSPECT? DETECT? TEST?**	**REWORK? WASTE? SCRAP?**

The QI Connect-the-Dots Book

Purpose

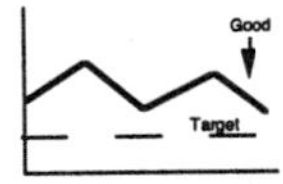

Define specific ways to measure the customer's requirements and to predict the stability and capability of the process.

Indicators

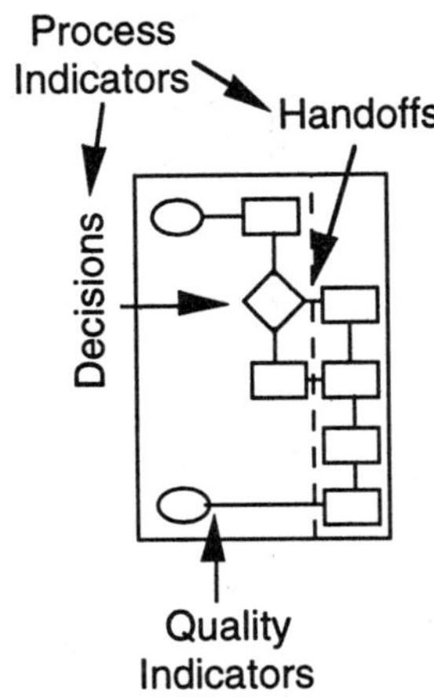

Goodness is uneventful. It does not flash, it glows.
  - David Grayson

A stone thrown at the right time is better than gold given at the wrong time.
  Persian Proverb

Men go shopping just as men go out fishing or hunting, to see how large a fish may be caught with the smallest hook.
  -Henry Ward Beecher

All problems invariably stem from failing to meet or exceed a customer's requirement. To begin to define the problem, you need to identify your customer's needs and a way to measure them *over time*--by hour, day, week, or month.

Quality indicators measure how well the product or service meets the customer's requirements. Process indicators, strategically positioned at critical hand off points in the process, provide an early warning system. For each "quality indicator" there should be one or more "in process" indicators that can predict whether we will deliver what they require.

Requirement	Indicators		Period
	Quality or Process		
Better	Number of defects Percent defective (number of defects/total)		minute hour day
Faster	# or % of commitments missed time in minutes, hours, days		week month
Cheaper	cost per unit cost of waste or rework		shift project

There are usually only a few key customer requirements for any product or service. Using the form on the next page, identify your main supplier, customer, the product or service used, and the process that creates it. Begin by identifying your requirements of the supplier. Then, identify your customer's requirements for the product or service. What do they want in terms of good, fast, and cheap. Then, based on your customer's needs, identify how you can measure it with defects, time, or cost. Finally, identify how often you will measure: by minute, hour, day, week, or month.

Step 2
Identify the Indicators

Main: **Customer** ________________

Supplier ________________

Product or Service ________________

Process ________________

Type	Requirement	Measurement	Period
Better	Do it right	<u>Number of defects</u> (inaccuracies, errors) <u>Number defective</u> (scrap, rework, complaints) <u>Percent defective</u> (number defective/total)	
Faster	Meet your commitments	<u>Number or percent commitments missed</u> <u>Time to deliver product or service in hours, days, etc.</u>	
	Fix it fast	<u>Wait or idle time</u>	
Cheaper		<u>Cost of rework</u> <u>Cost of waste</u> <u>Cost per unit</u>	

Example

	Requirement	Measurement	Period
Better	Do it right	Number of incorrect customer orders	Week
Faster	When I want it	Number of missed delivery commitments	Day
Cheap	Reduce my costs	Cost of correcting inaccurate orders	Day

Step 2 - Identify the (Quality) Indicators

Quality Indicators

To each thing belongs its measure.
- Pindar

Examples

Timeliness is best in all matters
- Hesiod

Most measures of customer requirements can be easily expressed as a line graph (showing periodic variation over time) or a histogram (pg. 42-43) showing a snapshot of overall performance:

Requirement:
I want an accurate paycheck

Measurement:
Paycheck errors

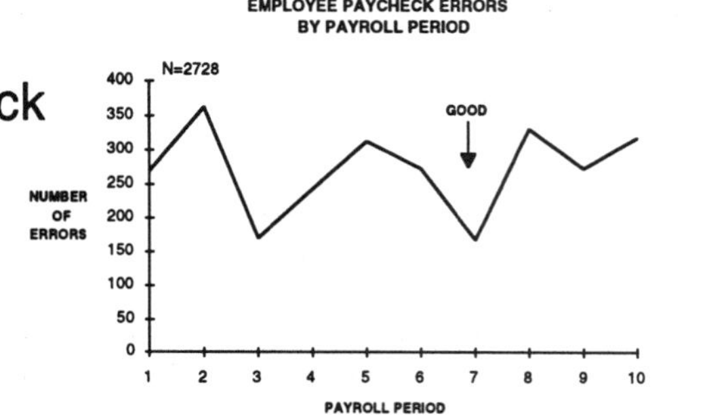

Requirement:
Don't waste my time
Do it in an hour or less

Measurement:
Time to install service

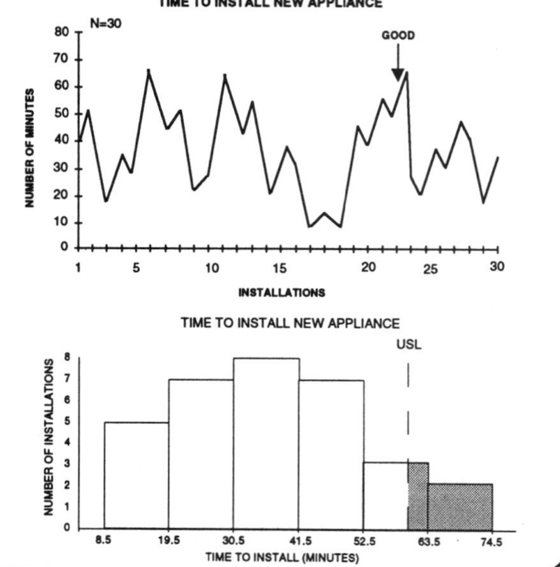

Requirement:
Be efficient
Don't waste food

Measurement:
Cost of
 food spoilage

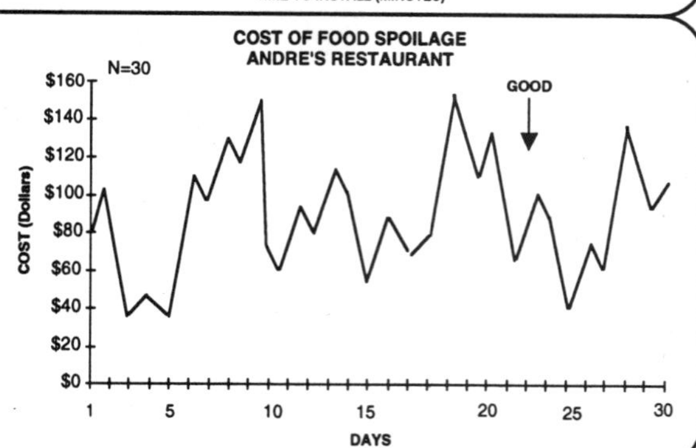

Step 2 - Identify the (Quality) Indicators

Quality Indicators

TIP: Indicators (of the customer's requirements) are measured <u>after</u> the product or service is delivered.

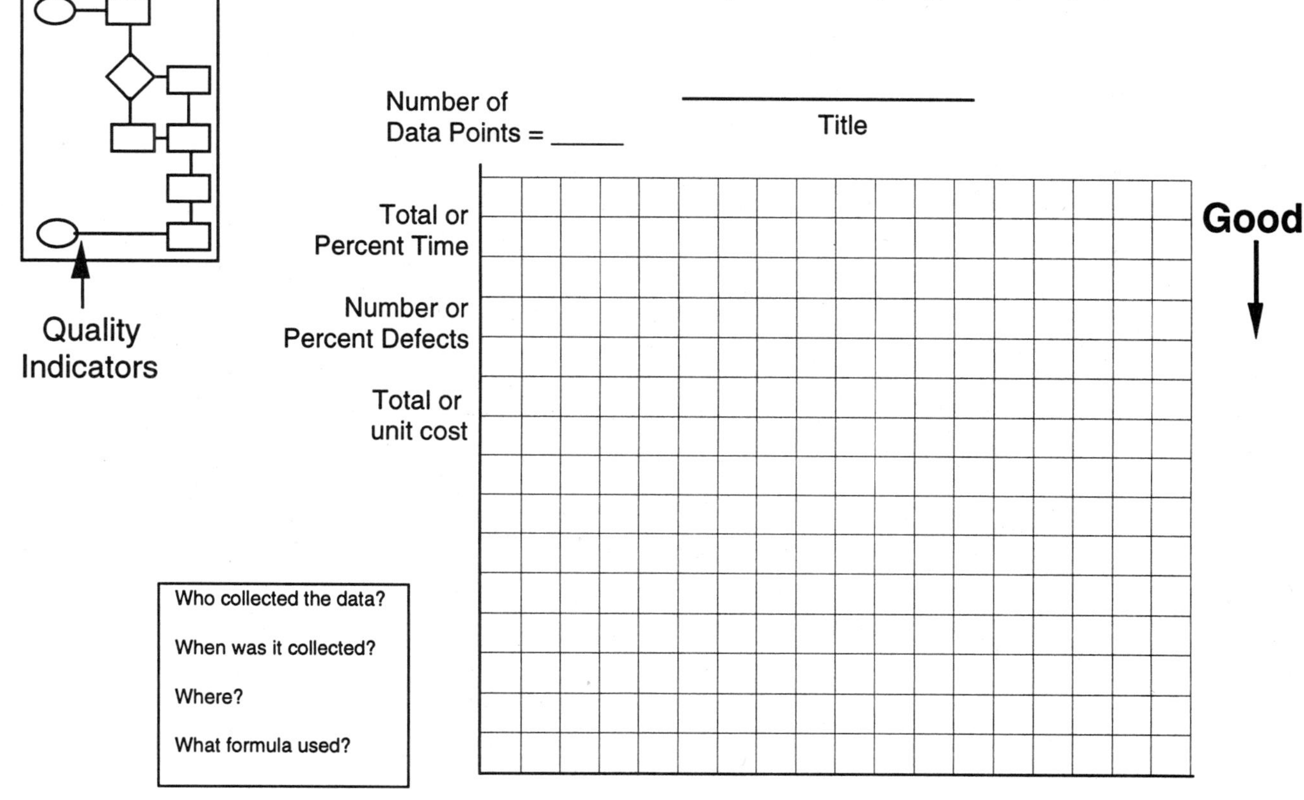

Step 2 - Identify the (Process) Indicators

Process Indicators

The measurements of customer requirements usually occur after the end product or service is delivered. To ensure that customers get what they want, we have to set up a system of early warning indicators that will predict whether or not the process will deliver what the customers want. Like the quality indicators, these predictive indicators will need to measure defects, time, and cost <u>inside the process</u>.

Where to Measure

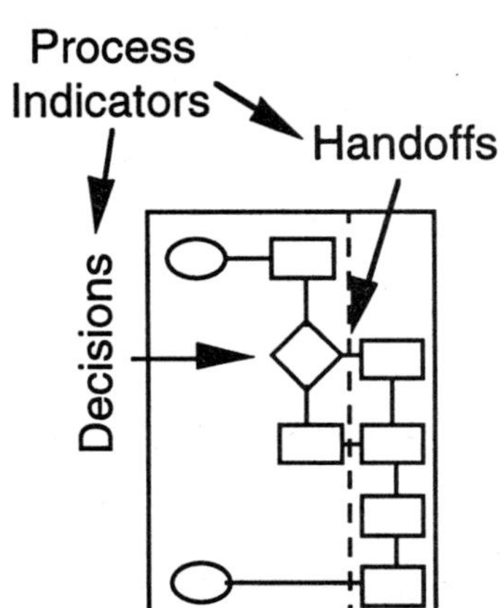

Often, the easiest points to measure **process indicators** are at the key handoffs (where the dashed lines intersect on the flowchart) or decision points . If, for example, you were trying to <u>predict </u>how long it would take to get to work, the number of red lights or average highway speed could <u>predict</u> the total commute time. Measure **quality indicators** (e.g., total commute time) <u>after</u> the product or service has been delivered. Looking back at the main flowchart, at what points could you most easily take measurements that would predict whether the process will be able to deliver what the customers want?

Examples

"Quality" Requirements Indicator	"Process" Early Warning Indicator
Percent defective	Amount of rework per step Number of defects per step
Missed Commitments	Time per process step Delay (idle and rework time)
Value	Cost of waste and rework
Paycheck errors	Timesheet errors % timesheets late
Appliance installation time	Old appliance removal time
Cost of food spoilage	Number of customers Perishable food ordered

Process Indicators

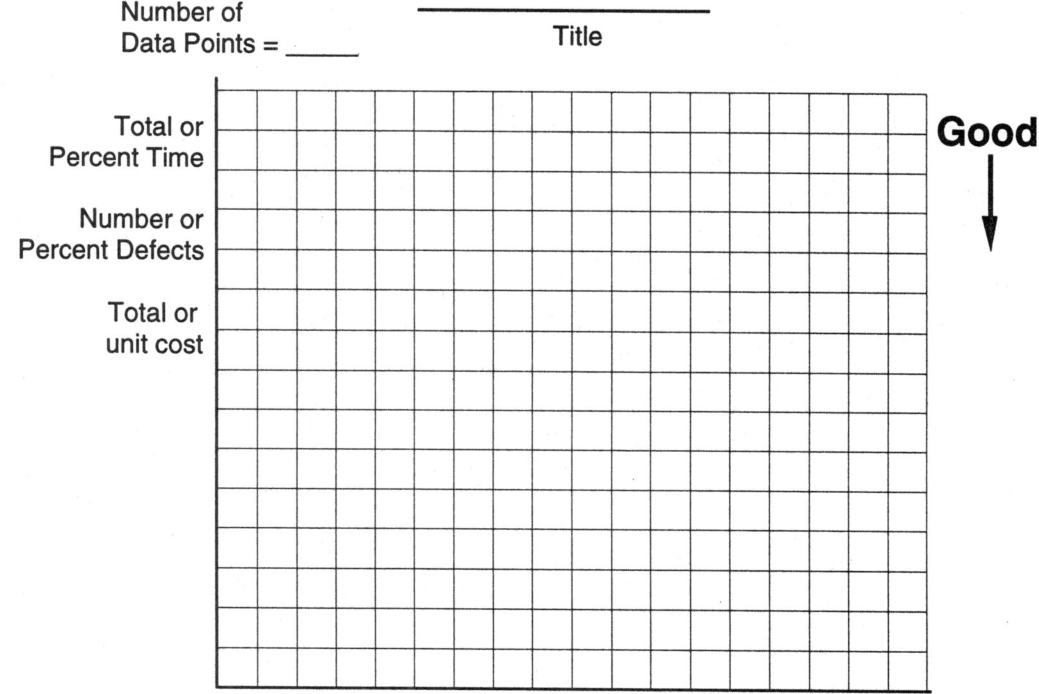

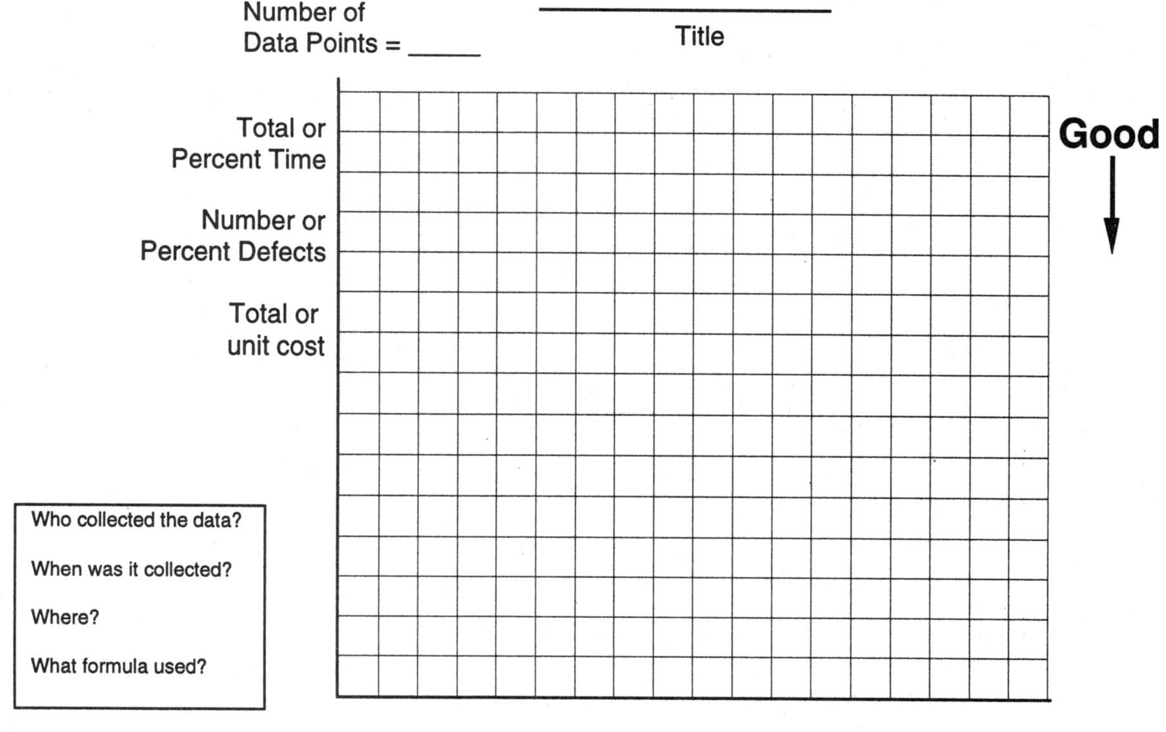

Step 3
Implement the Process

Purpose	Establish the baseline process and begin collecting the quality indicators

Planning

Like crops in a garden, most processes will require a careful plan to ensure they take root and flourish. To establish your process and to begin measuring will require a detailed plan of action.

A good implementation plan will identify:

> • **What** changes are needed in the people, process, machines, materials, and the environment
>
> • **How** to make the changes
>
> • **Who** will do them
>
> • **When** they will be started and completed
>
> • **How** they will be measured

Step 3
Implement the Process

ACTION PLAN

WHAT? (Changes)	HOW? (Action)	WHO?	WHEN? Start Complete		MEASURE? (Results)
Develop Buy-In	Ensure Sponsorship Communicate				
People	Training				
Process	Implement Measure Monitor				
Machines (Computers, vehicles, etc.)					
Materials (Forms & Supplies)					
Environ-ment					

Step 4 - Stabilize the Process
Understanding Stability

Stability

A stable process produces <u>predictable results consistently</u>. Stability can be easily determined from control charts. The upper control limit (UCL) and lower control limit (LCL) are <u>calculated</u> from the data.

Example

How long does it take you to commute to work each morning?

Stable = Predictable

Your Requirements
1. Get to work in 30 minutes or less.
2. Get to work safely (no faster than 15 minutes).

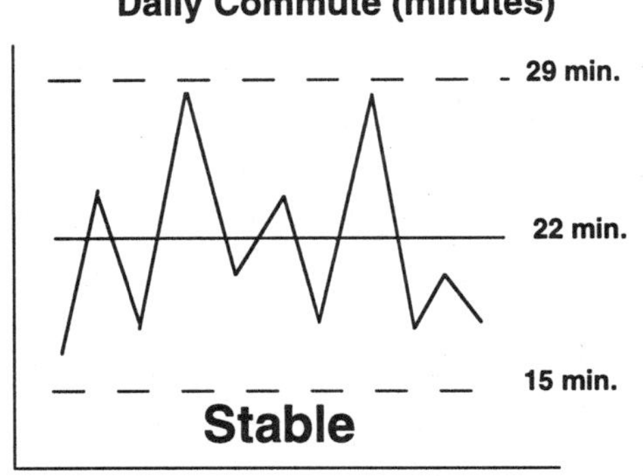

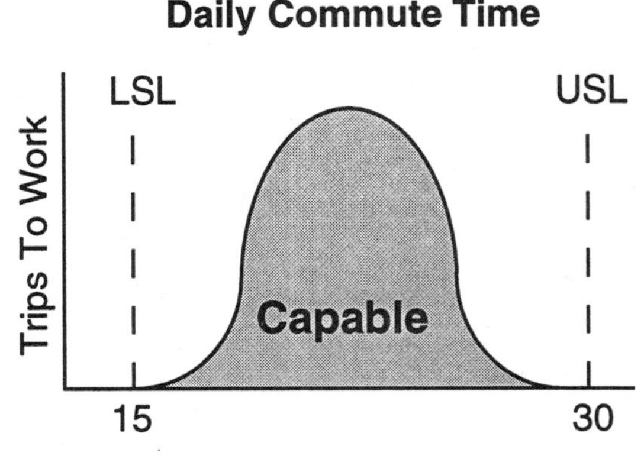

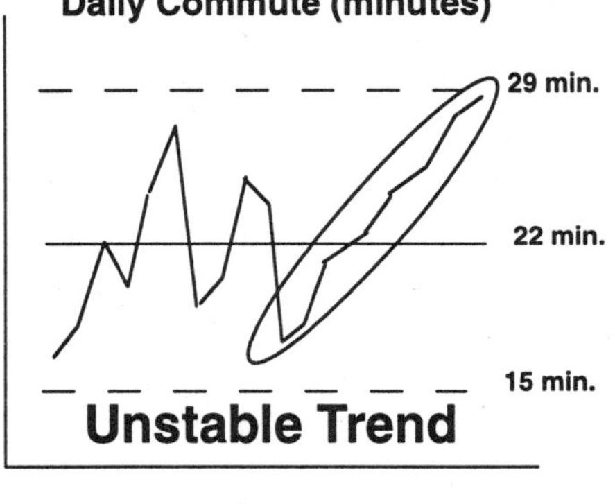

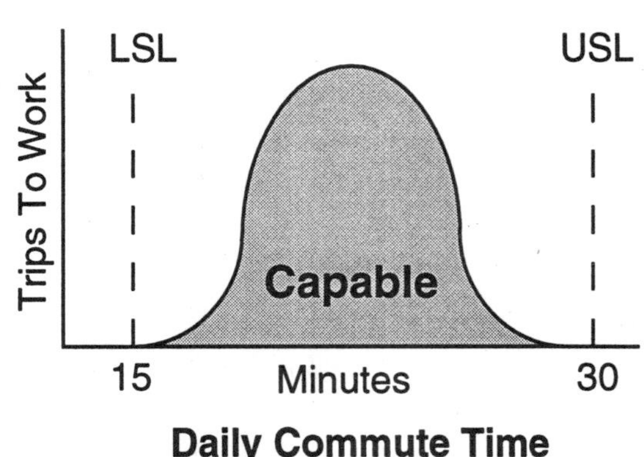

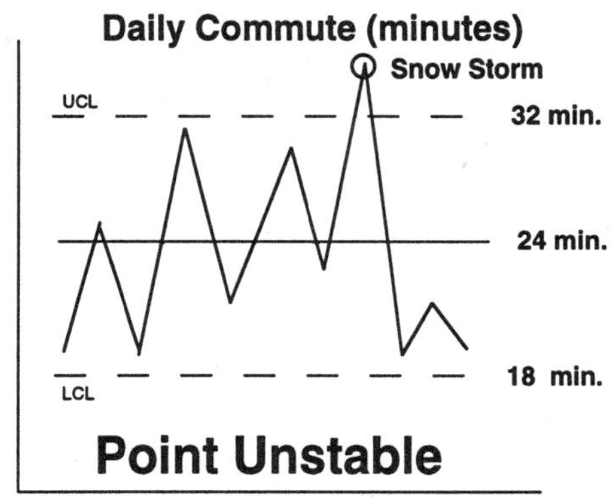

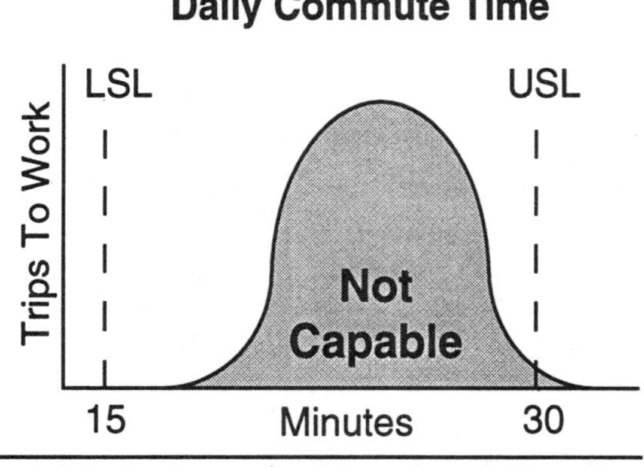

Stability and Capability

A process does not have to be stable to be capable of meeting the customer's requirements. Similarly, a stable process is not necessarily capable. A managed process must be both stable <u>and</u> capable. Interpreting stability with control charts and capability with histograms will be discussed in more detail on the following pages.

Step 4 - Stabilize the Process
Understanding Capability

Capability

A **capable** process <u>meets the customer's requirements 100% of the time</u>. The upper (USL) and lower (LSL) <u>specification</u> limits are determined from the customer's requirements.

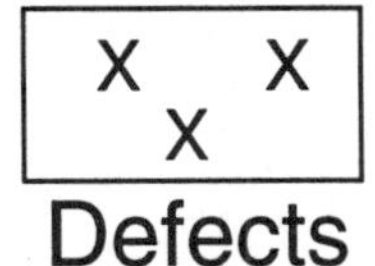

Defects

The <u>capability</u> of **counted** data (e.g., defects--indivisible integers only) is <u>zero defects</u>. Customers hate defects--outages (USL = LSL = 0).

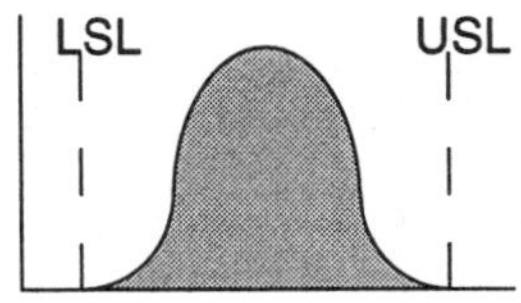

The <u>capability</u> of **measured** data (e.g., time, money, age, length, width, weight, etc.) is determined using the customer's specifications and a histogram (see below).

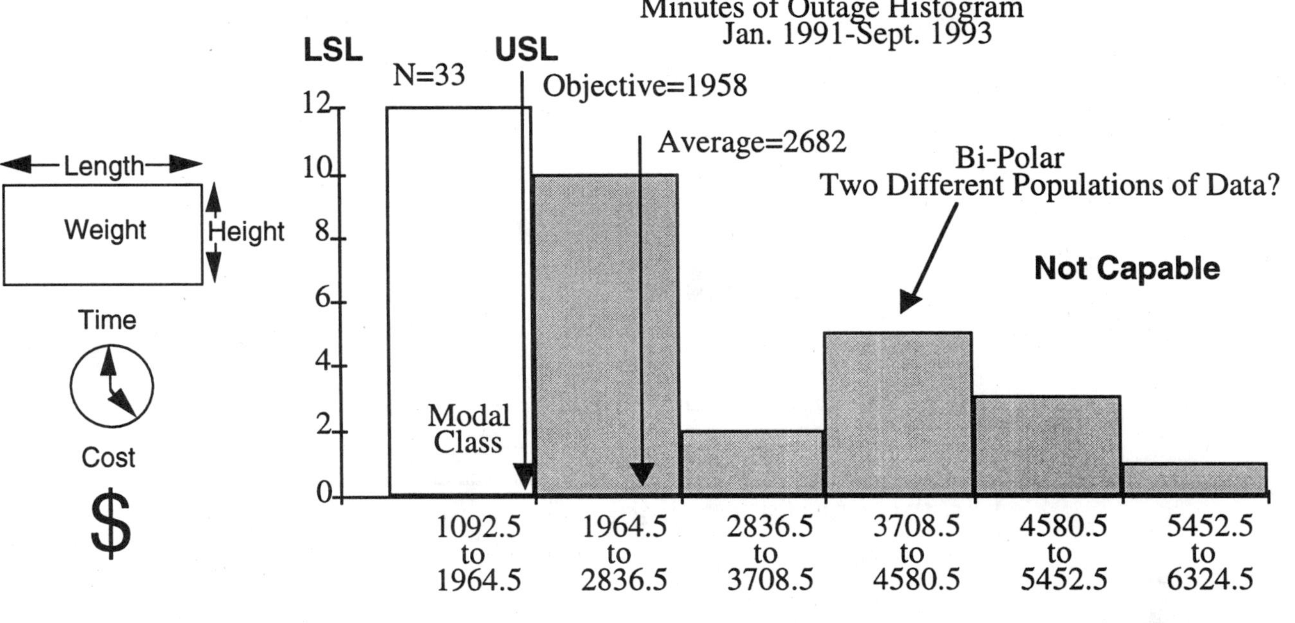

Problem Solving

Is the process capable? If not, what improvement activities are required to make the process both stable and capable?

Capable = Meets Customer Requirements 100% of the Time

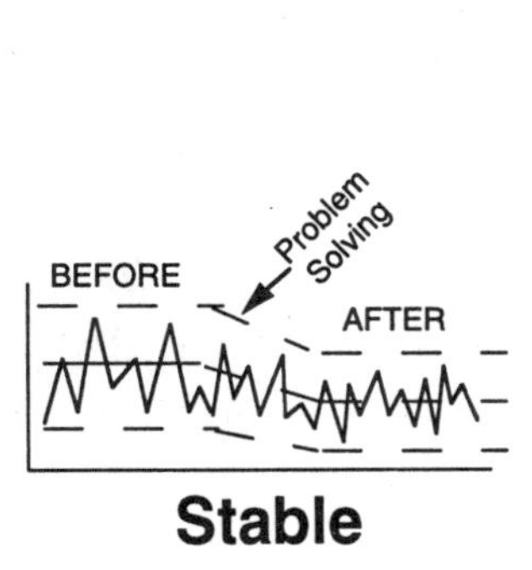

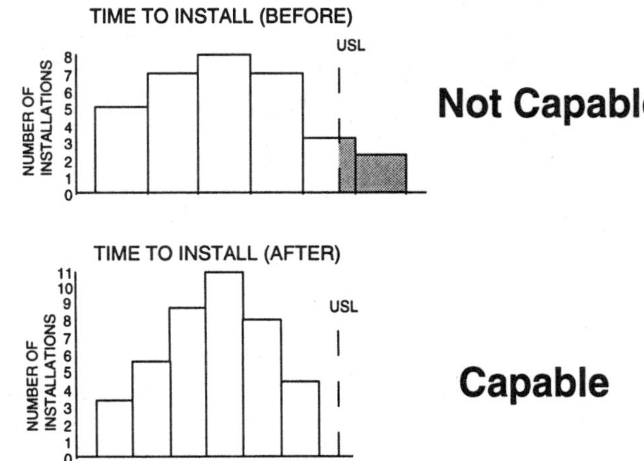

Step 4 - Check Stability
Sampling

Purpose	Minimize the cost of collecting the data.

Sampling

By a small sample may we judge the whole piece.
- Cervantes

To begin managing a process, you will need a minimum of 20 data points for each indicator. To collect data cost effectively, you will have to understand the basic concepts of sampling. If you are only producing ten widgets a day, then it is fairly easy to look at all of them (the *total population* of created widgets). If you produce 100,000 widgets a day, however, you will want to look at a small sample and *draw conclusions* about the entire population from the sample:

Lot or Total Population

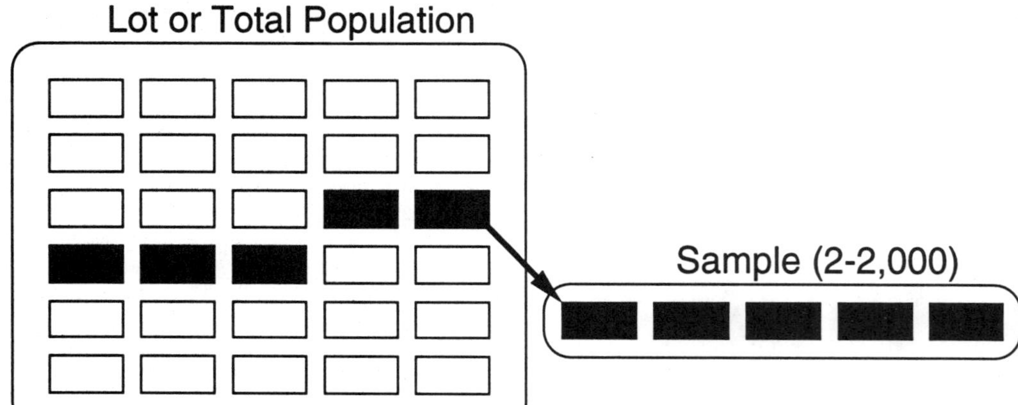

In the population, if there is a . . .	then . . .
large number that are measured	
manually	take a sample
mechanically	use the total population
small number	use the total population

Sample Size

When samples are taken, they should be the same size: If you check five widgets in this sample, then you will want to check five widgets *every* time you sample. If you use the total population, it can vary from period to period or be a constant size. In general, the sample size will vary based on: the acceptable level of quality desired, the size of the "lot" you're inspecting, the type of sampling done--single, double, multiple, and the level of inspection. Higher quality requires larger samples.

Step 4 - Check Stability
Interpreting The Indicators

| **Purpose** | Verify that the process system is stable and can predictably meet customer requirements |

Variation

You cannot step twice into the same river.
Heraclitus

A stable process produces <u>predictable results</u>. Understanding variation helps us learn how to predict the performance of any process. To ensure that the process is stable (i.e., predictable) we need to develop "run" or "control" charts of our indicators.

How can you tell if a process is stable? Processes are never perfect. *Common* and *special causes* of variation make the process perform differently in different situations. Getting from your home to school or work takes varying amounts of time because of traffic or transportation delays. These are <u>common causes</u> of variation; they exist every day. A blizzard, a traffic accident, a chemical spill, or other freak occurrence that causes major delays would be a <u>special cause</u> of variation.

In the 1920s, Dr. Shewhart, at Bell Labs, developed ways to evaluate whether the data on a line graph is common cause or special cause variation. Using 20-30 data points, you can determine how stable and predictable the process is. Using simple equations, you can calculate the average (center line), and the upper and lower "control limits" from the data. 99% of all *expected* (i.e., common cause variation) should lie between these two limits. Control limits are not to be confused with specification limits. Specification limits are defined by the customer. Control limits show what the process can deliver.

Example

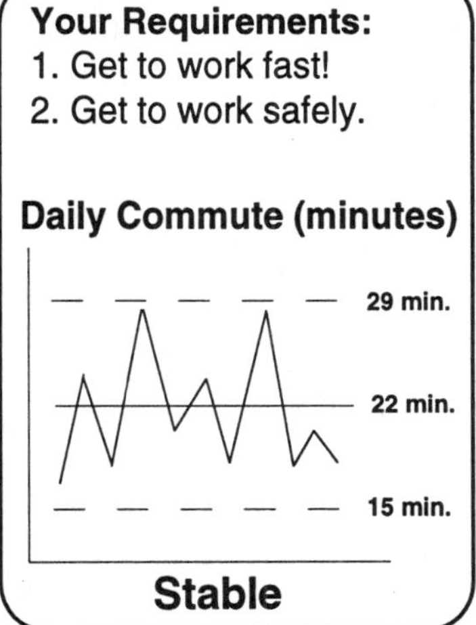

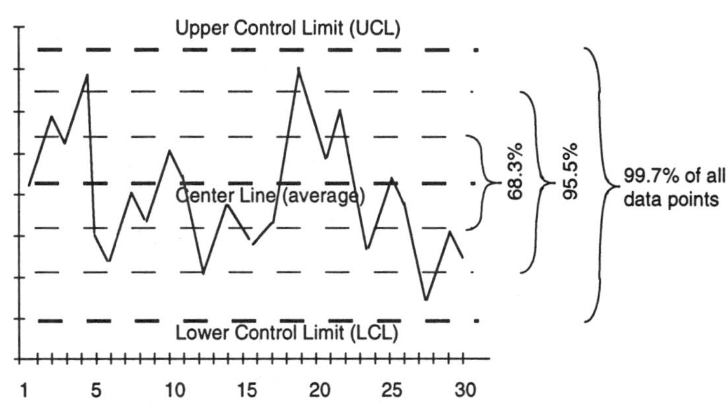

Step 4 - Check Stability
Interpreting The Indicators

Special Cause Variation

Processes that are "out of control" need to be stabilized before they can be improved using the problem-solving process. Special causes, require immediate cause-effect analysis to eliminate the special cause of variation.

Evaluating Stability

The following diagram will help you evaluate stability in any control chart. Unstable conditions can be any of the following:

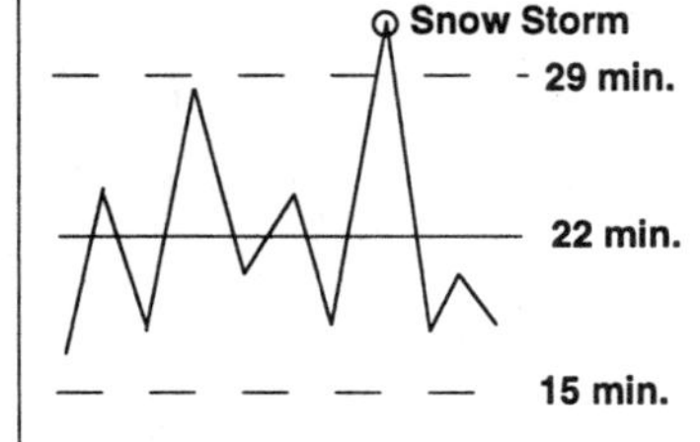

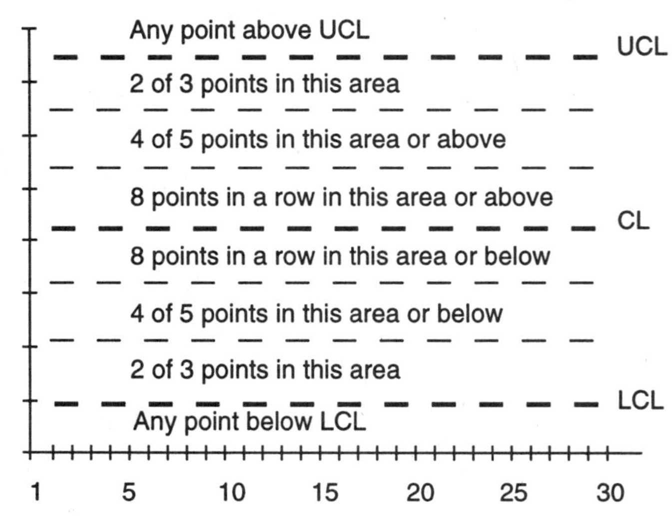

Points and Runs

Any point outside the upper or lower control limits is a clear example of a special cause. The other forms of special cause variation are called "runs." Trends, cycling up and down, or "hugging" the center line or limits are special forms of a run.

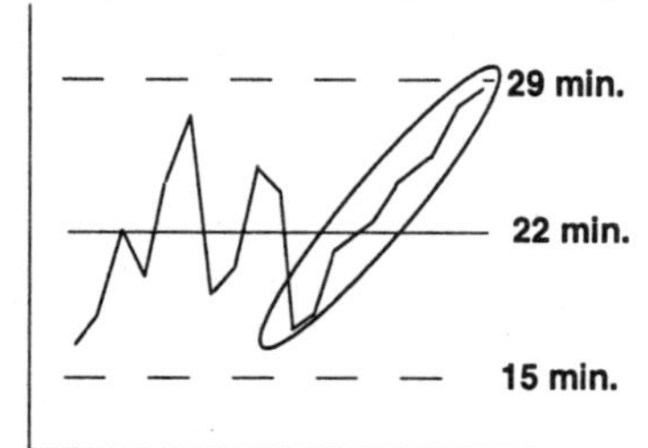

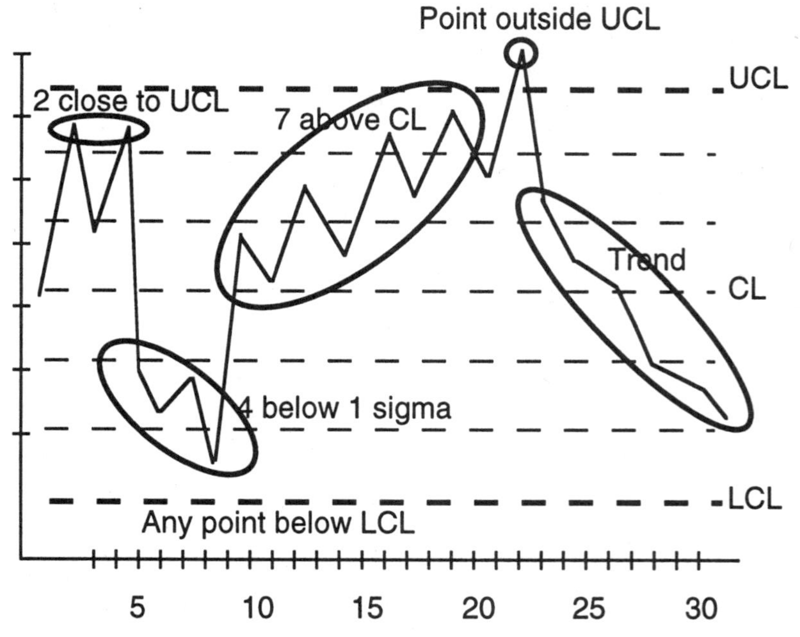

Types of Charts

Next, you have to know what to collect about each widget. Do you need to know: how long it takes to deliver a product or service, the number of defects per product, or the cost of waste or rework? Time, cost, length, and weight are known as *variable* data. Counting the number of defects or defective items gives *attribute* data. The type of data (attribute or variable) and the size of the sample taken (1, 2-10, or total) will determine the type of graph used to measure the process.

Choosing a Chart

Type of data	Sample Size			
	1	Same Size	Varies	
Attribute data — $\boxed{X}$ = 1 Defective (pp. 36-37)		**np** (2-total)	**p** (total)	
Attribute data — $\boxed{X \quad X \atop X}$ = 3 Defects (pp. 38-39)		**c** (2-total)	**u** (total)	Type of Chart
Variable data — Length, Weight, Height, Time, Cost $ (pp. 40-41)	**XRs**	$\overline{X}$R (size 2-10)		
Y Axis	Number	Number	Percent	

Drawing the Chart

Regardless of sample size, each of these charts can be drawn as a *line graph.* **Tip:** If the math seems scary, start with line graphs or get the QI Macros for Microsoft® Excel.

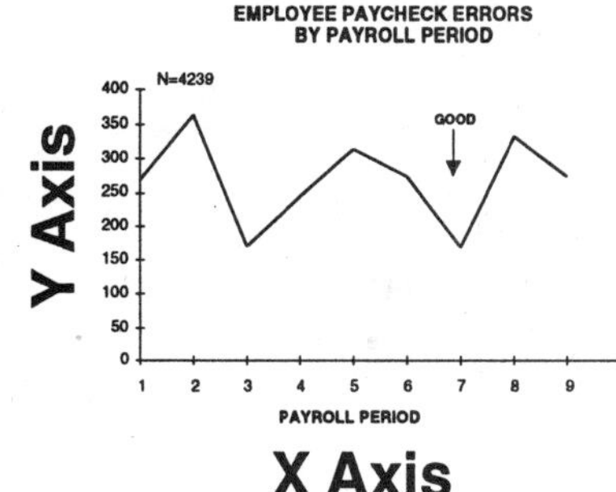

- The **X axis** (horizontal) shows how often the data is collected (daily, hourly, weekly, periodically).
- The **Y axis** (vertical) shows:
    - the number or the percent defective (c, np, p, u)
    - the time, cost, length, weight, etc. (XR charts)

Then, based on the type of data and sample size, you can calculate the upper and lower control limits (UCL, LCL) and center line (CL) that will make it possible to evaluate process stability. The next few pages show how to calculate and interpret the limits.

Step 4 - Check Stability
np and p charts

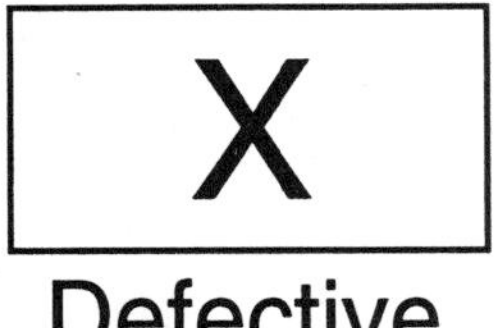

p and np Charts
(Attribute data)

The p and np charts will help you evaluate process stability when counting the number or fraction <u>defective</u>. Examples might include: the number of defective circuit boards, meals in a restaurant, teller interactions in a bank, invoices, or bills.

The np chart is useful when it's easy to count the number of defective items and the sample size is always the same. The p chart is used when the sample size varies: the total number of circuit boards, meals, or bills delivered varies from one sampling period to the next. In the p chart below, the number of defective paychecks varies with the number of employees in each pay period.

Paychecks - Fraction Defective

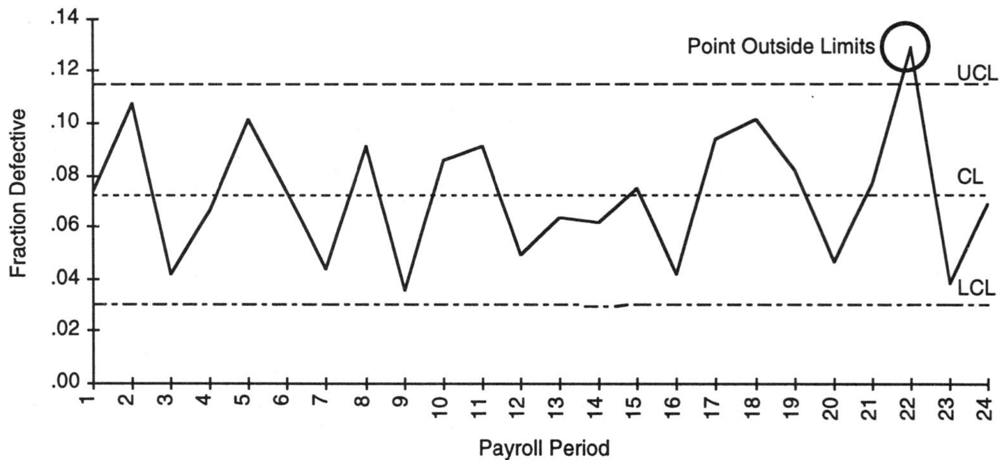

Stability

Given this information, we would want to investigate why the 22nd payroll period was "out of control." Otherwise, this chart, and therefore this process, look stable.

Capability

A fully capable process delivers <u>zero defects</u>. Although this may be difficult to achieve, it should still be our goal. Once we resolve the out-of-control point, we could use the quality problem solving process to begin to eliminate the common causes of defective paychecks. What are the most common types of paycheck errors? Why do they occur? What are the root causes of these paycheck errors?

Step 4 - Check Stability
np and p charts

X	= 1 Defective

____
Title

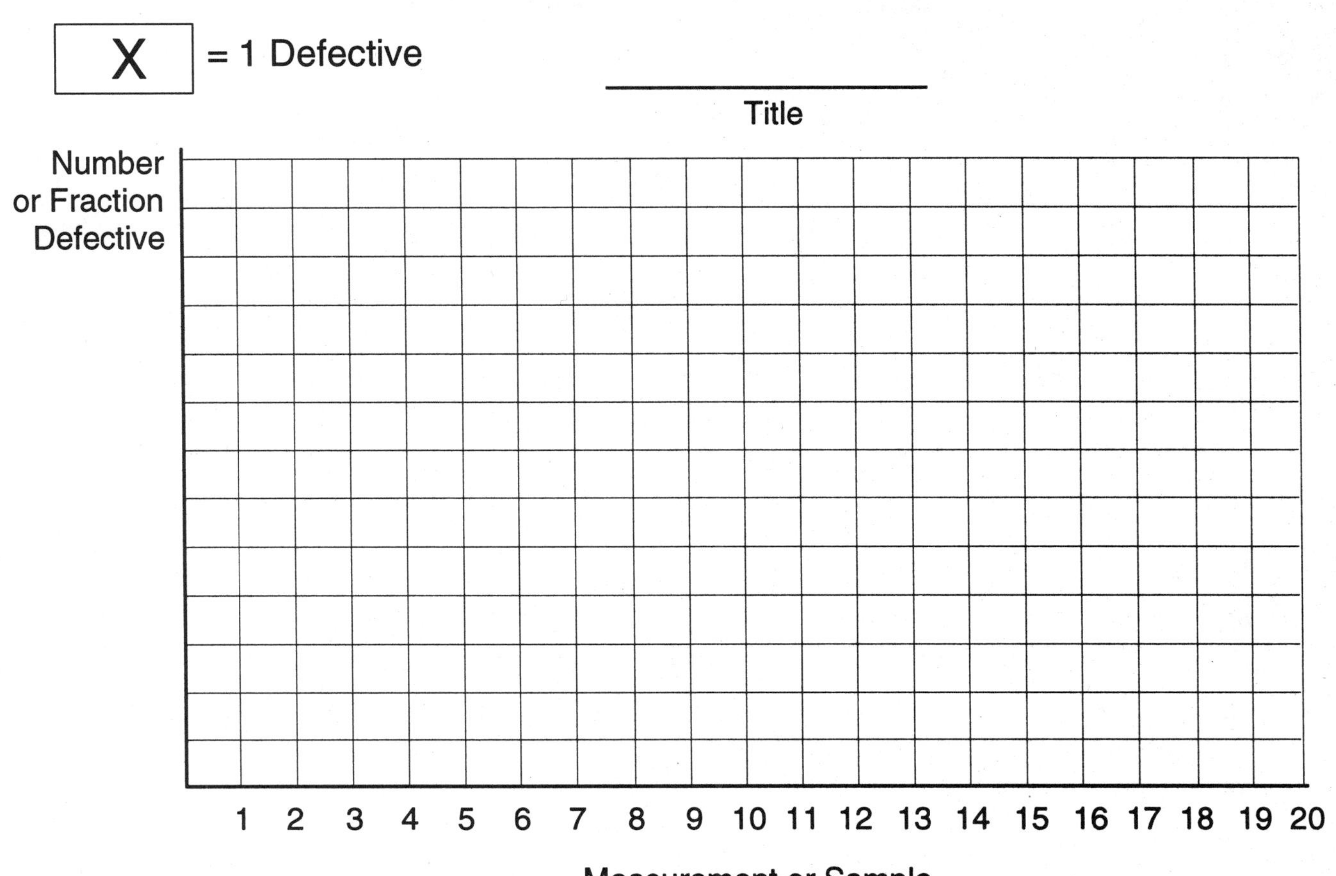

p

	1	2	3	4	5	6	7	8	9	10	11	12	13	14	15	16	17	18	19	20
Defective (p)																				
Sample Size																				
Fraction																				
UCL																				
LCL																				

np

	1	2	3	4	5	6	7	8	9	10	11	12	13	14	15	16	17	18	19	20
Defective (np)																				
Sample Size																				

p Chart

UCL: $\bar{p} + 3*\text{sqrt}(\bar{p}*(1-\bar{p})/n_i)$

CL: $\bar{p} = \Sigma p_i / \Sigma n_i$

LCL: $\bar{p} - 3*\text{sqrt}(\bar{p}*(1-\bar{p})/n_i)$

np Chart

$n\bar{p} + 3*\text{sqrt}(n\bar{p}*(1-n\bar{p}/n))$

$n\bar{p} = \Sigma np_i / k$

$n\bar{p} - 3*\text{sqrt}(n\bar{p}*(1-n\bar{p}/n))$

c and u Charts
(Attribute data)

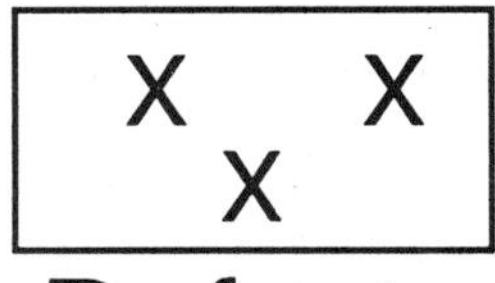

Defects

The c and u charts will help you evaluate process stability when there can be more than one defect per unit. Examples might include: the number of defective elements on a circuit board, the number of defects in a dining experience--order wrong, food too cold, check wrong, or the number of defects in bank statement, invoice, or bill. This chart is especially useful when you want to know how many defects there are not just how many defective items there are. It's one thing to know how many defective circuit boards, meals, statements, invoices, or bills there are; it is another thing to know how many defects were found in these defective items.

The c chart is useful when it's easy to count the number of defects and the sample size is always the same. The u chart is used when the sample size varies: the number of circuit boards, meals, or bills delivered each day varies. The c chart below shows the number of defects per day in a uniform sample.

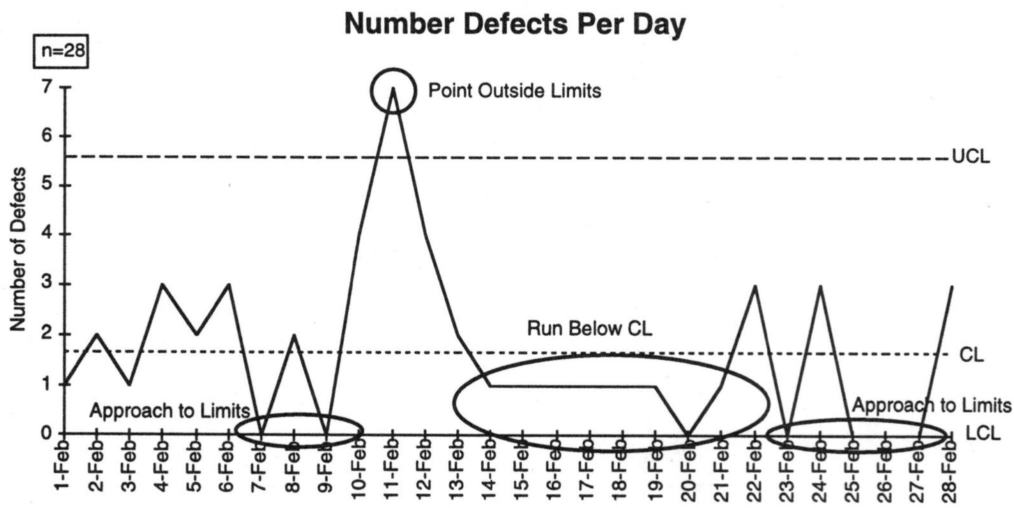

Stability

Given this information, we would want to investigate why February 11th was "out of control." We would also want to understand why we were able to keep the defects so far below average in the other circled areas. What did we do here that was so successful?

Capability

A fully capable process delivers <u>zero defects</u>.

Step 4 - Check Stability
c and u charts

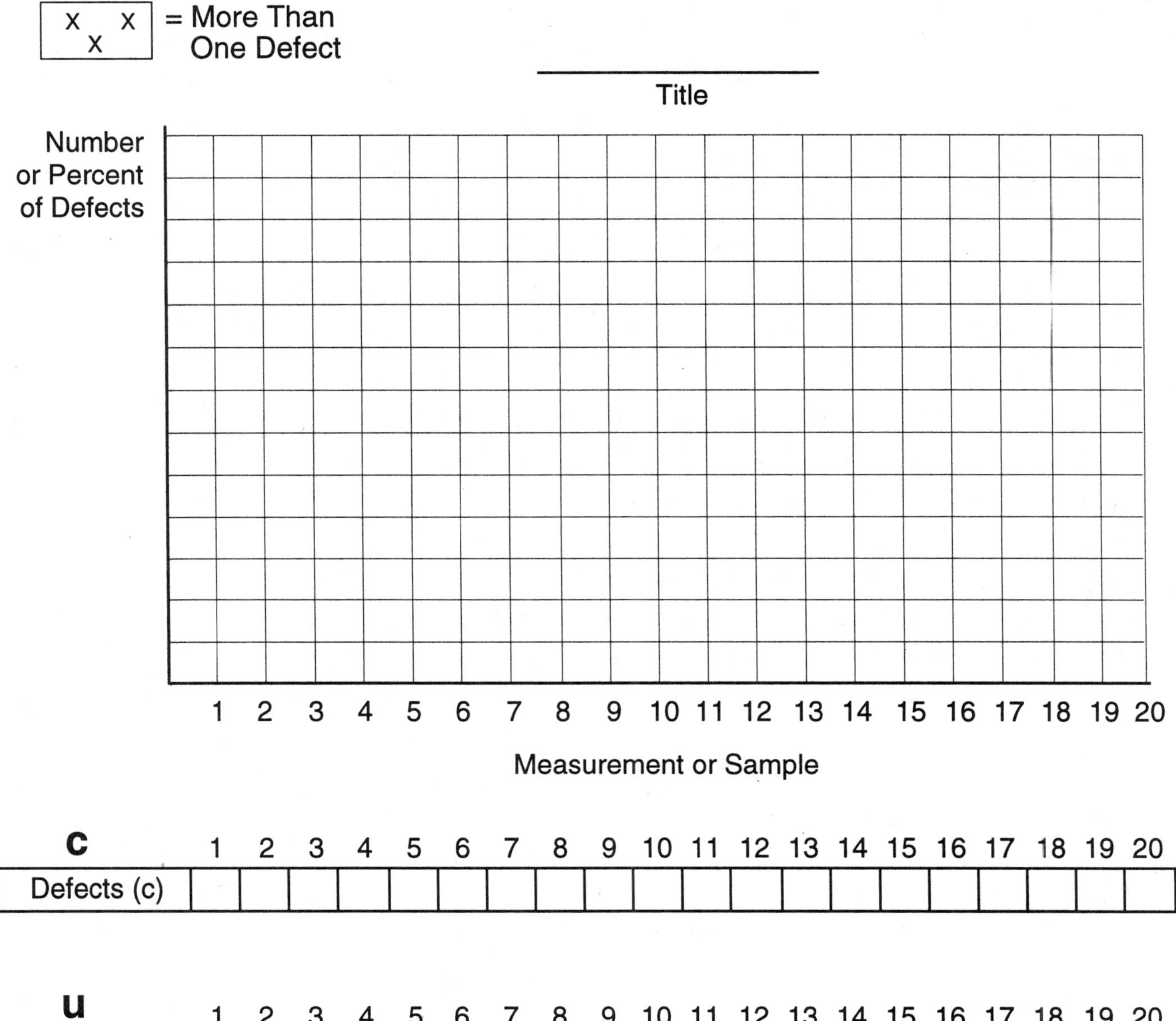

c

	1	2	3	4	5	6	7	8	9	10	11	12	13	14	15	16	17	18	19	20
Defects (c)																				

u

	1	2	3	4	5	6	7	8	9	10	11	12	13	14	15	16	17	18	19	20
Defects (u)																				
Sample Size (n)																				
Percent																				
UCL																				
LCL																				

C Chart		**U Chart**
UCL:	$\bar{c} + 3*sqrt(\bar{c})$	$\bar{u} + 3*sqrt(\bar{u}/n_i)$
CL:	$\bar{c} = \sum c_i/n$	$\bar{u} = \sum u_i/\sum n_i$
LCL:	$\bar{c} - 3*sqrt(\bar{c})$	$\bar{u} - 3*sqrt(\bar{u}/n_i)$

Step 4 - Check Stability
XR Charts

XR Chart
(Variable data Sample Size=5)

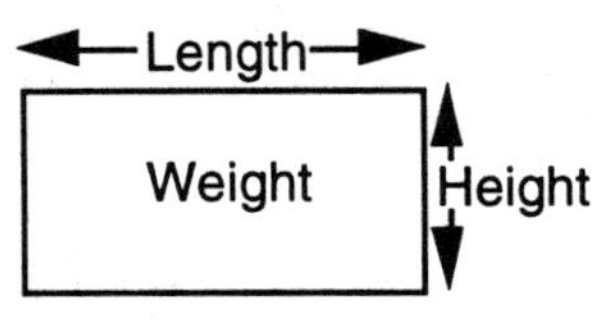

The $\overline{X}R$ chart can help you evaluate the cycle time for almost any process: making a widget, answering a customer call, seating a customer, delivering a pizza, or servicing an appliance. This chart is especially useful when you do this many times a day. Collecting the data could be expensive if you measured every time you did it. Using a small sample (typically five), however, you can effectively measure and evaluate the process.

$\overline{X}R$		R Chart	(For Sample Size=5)	X Chart
	UCL:	$2.114 * \overline{R}$		$\overline{X} + .577 * \overline{R}$
	CL:	$\overline{R} = \Sigma R_i / k$	$R = Max(X_i) - Min(X_i)$	$\overline{\overline{X}} = \Sigma \overline{X}_i / k$
	LCL:	0		$\overline{\overline{X}} - .577 * \overline{R}$

$\overline{X}R$

	1	2	3	4	5	6	7	8	9	10	11	12	13	14	15	16	17	18	19	20
Sample 1																				
Sample 2																				
Sample 3																				
Sample 4																				
Sample 5																				
Total																				
Average ($\overline{X}$)																				
Range (R)																				

XRs Chart
(Variable Data, Sample Size=1)

The XRs chart can help you evaluate a process when there is only one sample and they are farther apart: monthly postage expense, time to write a computer program, and so on.

XRs		R Chart	X Chart
	UCL:	$3.268 * \overline{R}$	$\overline{X} + 2.660\overline{R}$
	CL:	$\overline{R} = \Sigma R_i / (k-1)$ $R = abs(X_i - X_{i-1})$	$\overline{X} = \Sigma X_i / k$
	LCL:	0	$\overline{X} - 2.660\overline{R}$

XRs

	1	2	3	4	5	6	7	8	9	10	11	12	13	14	15	16	17	18	19	20
X Value																				
Range (R)																				

For both charts, you calculate, plot, and evaluate the <u>range chart first</u>. If it is "out of control," so is the process. If the range chart looks okay, then calculate, plot, and evaluate the X chart.

Step 4 - Check Stability
XR Charts

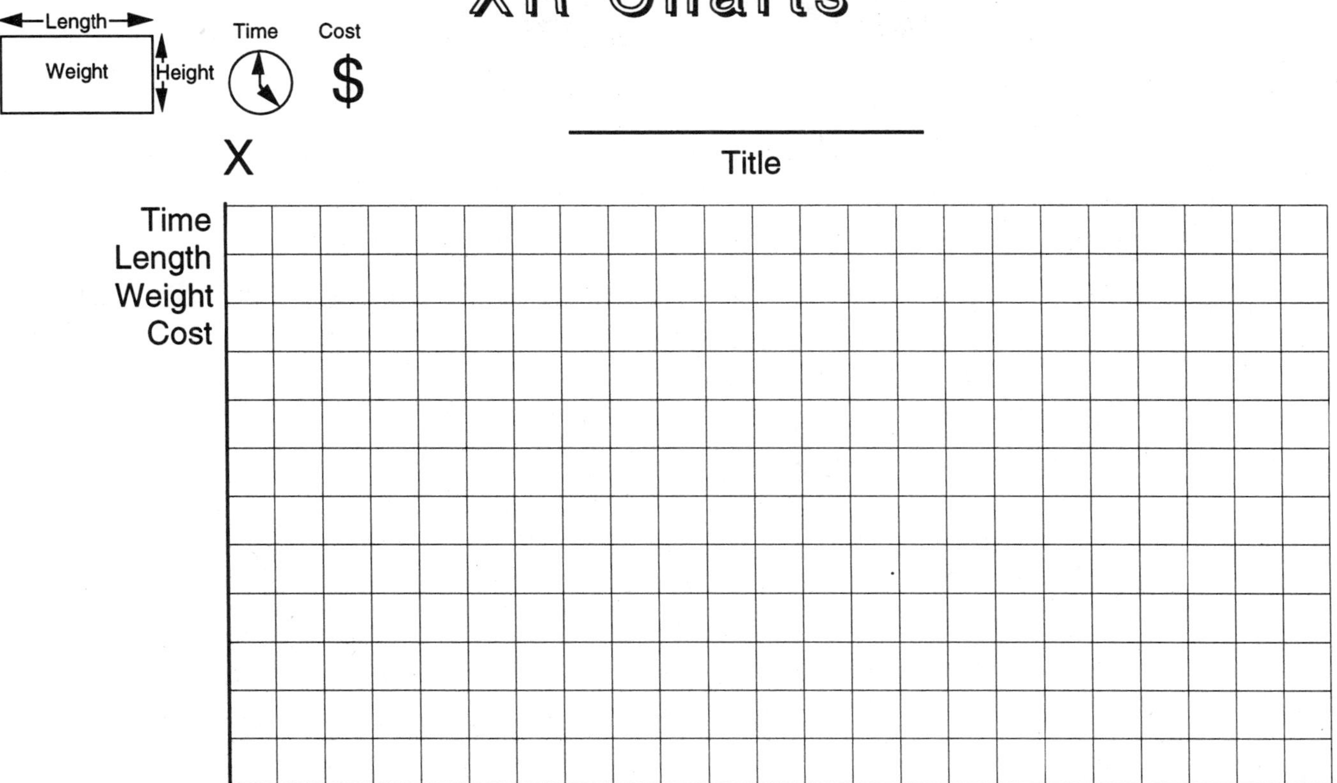

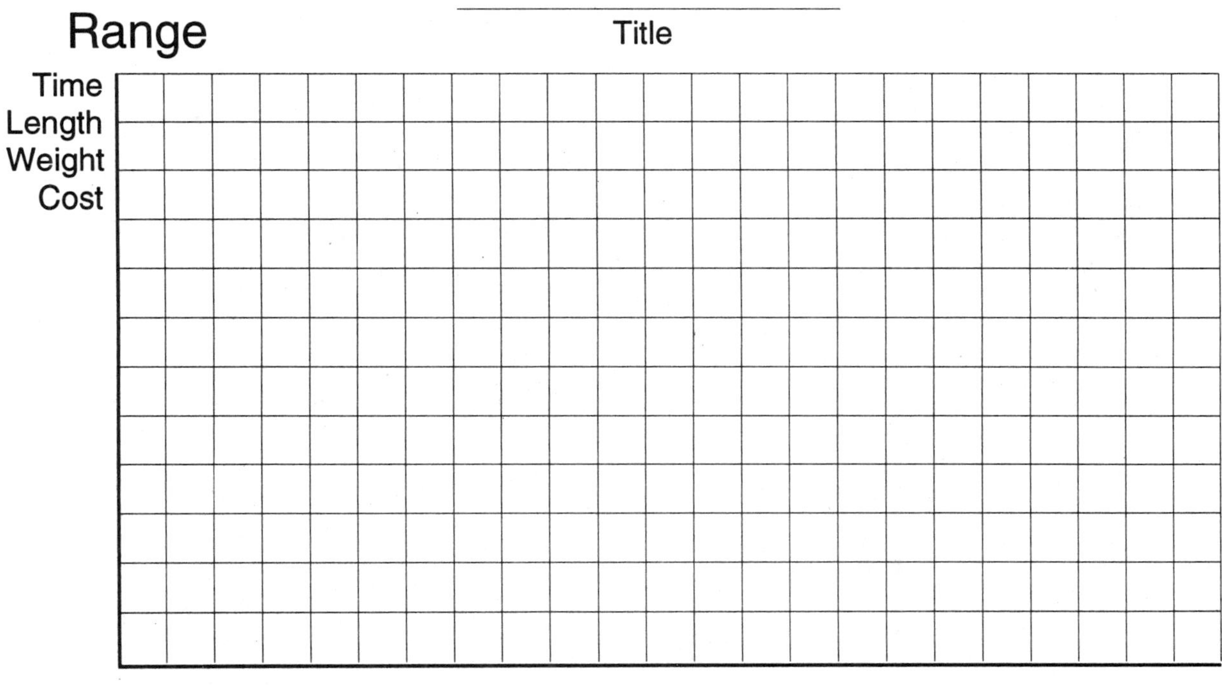

Step 4 - Check Capability
Using Histograms

Purpose	Verify that the process system is capable.

A **capable** process meets the customer's expectations 100% of the time.

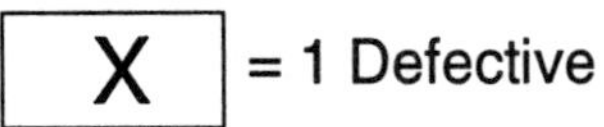

For **counted** data (e.g., defects), the process is capable when there are zero defects. (Customers want defect-free products.)

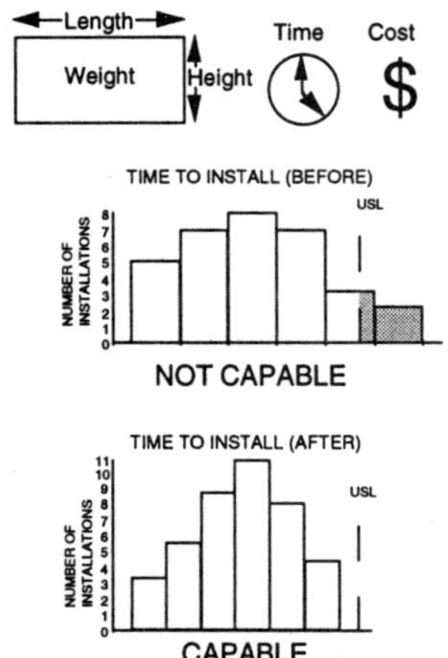

The capability of **measured** data (i.e., money, time, age, length, width, weight, etc.) is determined using a histogram.

Histograms show the spread, or dispersion, of data. The histogram uses variable (i.e., measured) data to determine process capability. The customer's upper (USL) and lower specification limits (LSL) determine process capability. When all of the data falls between the upper and lower specification limits, the process is capable.

Constructing a Histogram

Histograms are constructed as follows:

n	k
20-50	6
51-100	7
101-200	8
201-500	9
501-1000	10
1000+	11-20

mean = $\Sigma x/n$

median = middle x

mode = most frequent value

Step	Activity
1	Arrange the numbers in ascending order
2	Set n = number of data points (typically 30 or more)
3	Compute the range: R = maximum - minimum
4	Select the number of columns or classes (k), typically 6.
5	Compute width of columns: w = R/k
6	Establish the starting point = minimum - unit/2. The unit of measurement is typically 1, 0.1, or 0.01. Subtracting 1/2 a unit ensures that all points will easily fall within a given column and not on the dividing line. Subsequent points = current point + width (w).
7.	Tally the number of points that fall in each column of the histogram using the matrix on the next page.
8.	Draw the histogram using the tallies for each column
9.	Evaluate capability given customer specifications.

The QI Connect-the-Dots Book

Step 4 - Check Capability
Drawing a Histogram

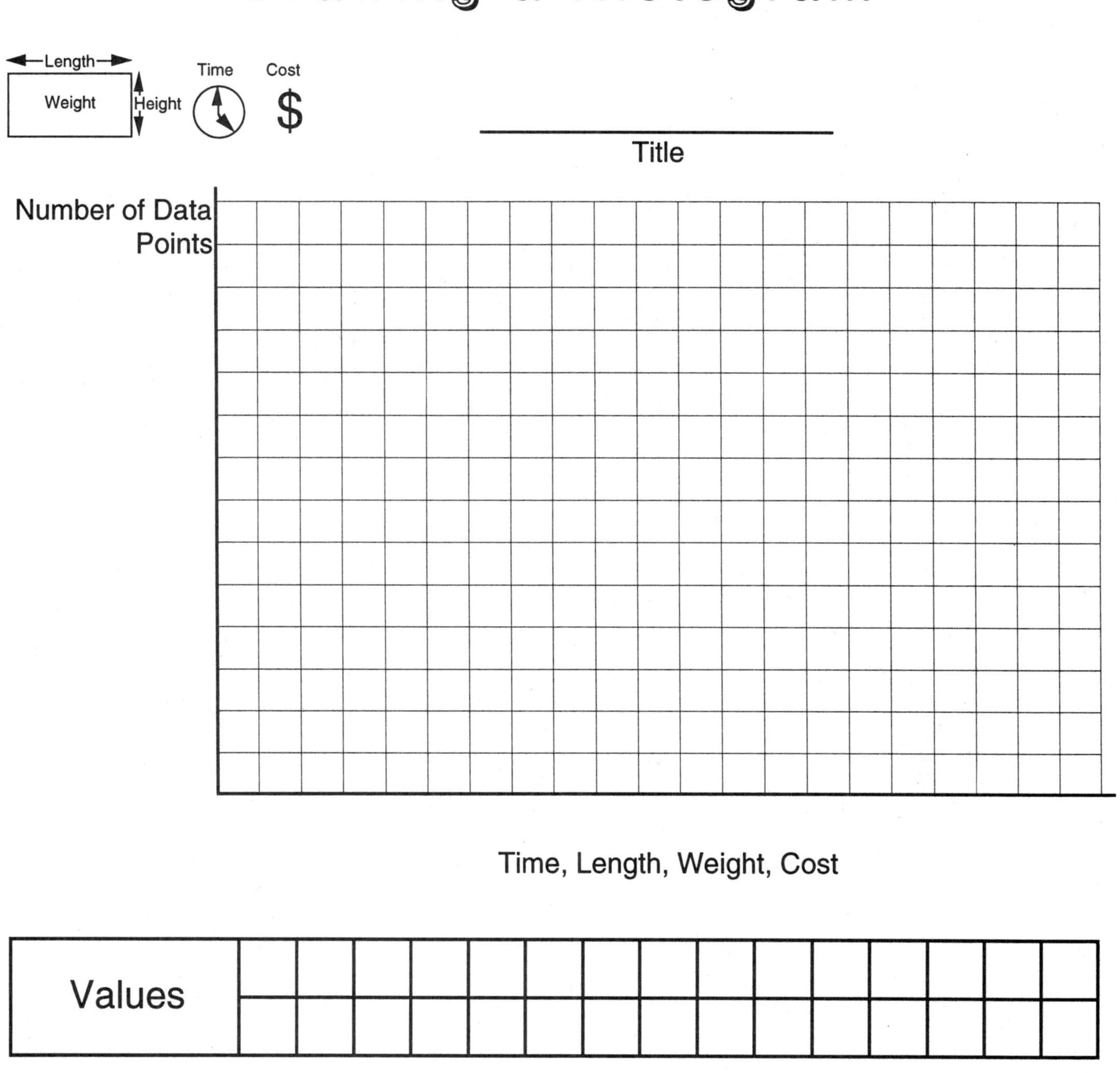

___ Range = Maximum - Minimum
___ Width of Bars = Range/6
___ Starting point = Minimum - Unit/2
___ End Bar 1 = Starting Point + Width
___ End Bar 2 = End Bar 1 + Width, etc.

. . .

Number per Column

1	2	3	4	5	6

Step 5 - Act To Improve
Root Cause Analysis

Purpose

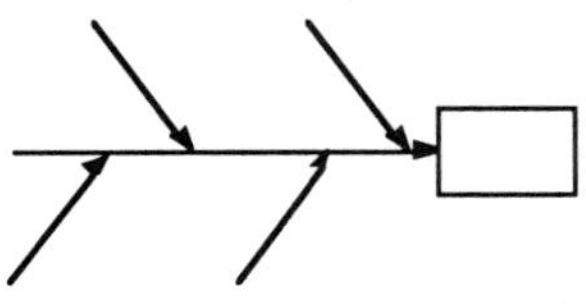

Improve process stability

If the process is <u>not stable</u>, we can use the quality improvement problem solving tools, especially the Ishikawa Diagram, to identify the root causes of the instability, remove them, and make the process stable, repeatable, and predictable.

Special Cause Analysis

For every thousand hacking at the leaves of evil, there is one striking at the root.
  -Thoreau

Every why has a wherefore.
  - Shakespeare

A good garden may have some weeds.
  Thomas Fuller, M.D.

1. To identify root causes, use the fishbone or Ishikawa diagram. Put a problem statement about the special cause of variation in the head of the fish and the major causes at the end of the major bones. Major causes include:

   - Processes, machines, materials, measurement, people, environment
   - Steps of a process (step1, step2, etc.)
   - Whatever makes sense

2. Begin with the most likely main cause.

3. For each cause, ask "Why?" up to five times.

4. Circle one-to-five <u>root</u> causes (end of "why" chain)

5. Verify the root causes with data (Pareto, Scatter)

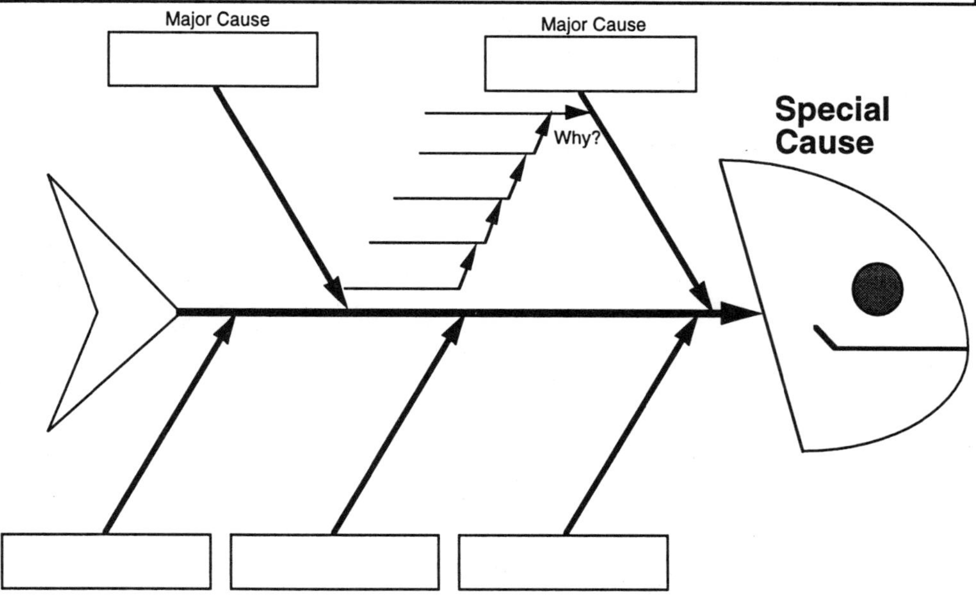

Step 5 - Act To Improve
Problem Solving

Purpose	Improve process capability

Common Cause Analysis

If the process is <u>not capable</u> of meeting the customer's needs and expectations, then we can use the quality problem solving process to begin to analyze and remove the so called "common causes" of variation in the process.

The problem-solving process follows the FISH cycle step-by-step to ensure continuous, never-ending improvement:

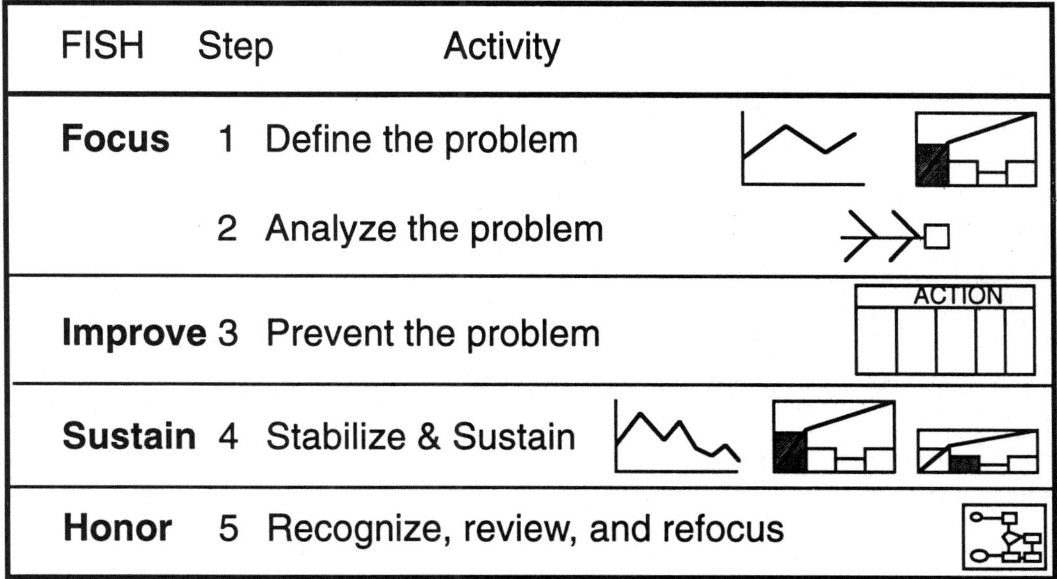

FISH	Step	Activity
Focus	1	Define the problem
	2	Analyze the problem
Improve	3	Prevent the problem
Sustain	4	Stabilize & Sustain
Honor	5	Recognize, review, and refocus

> For more detail, get the
> *QI Coloring Book*®
> *Problem Solving Made Easy*

Stability and Capability *after* Problem Solving

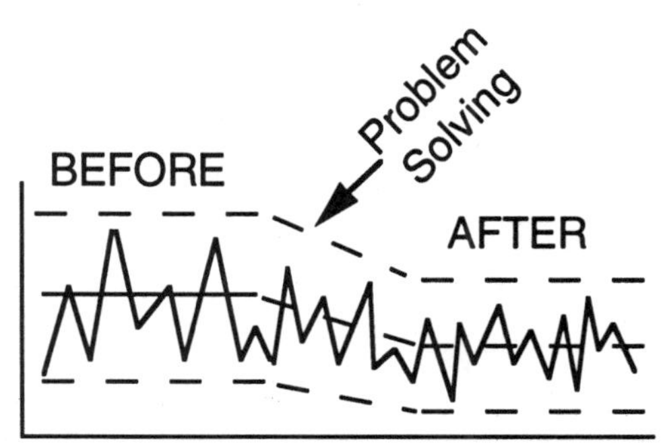

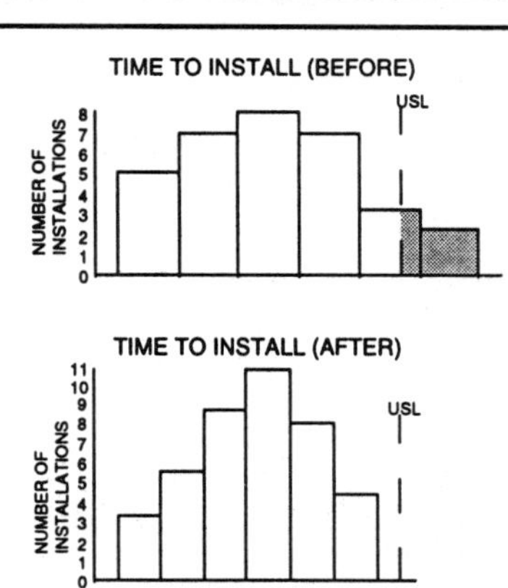

Step 5 - Act To Improve
Multiply The Gains

Purpose

To get the maximum benefit from having defined and managed this process, we need to get this process into the hands of everyone who can use it.

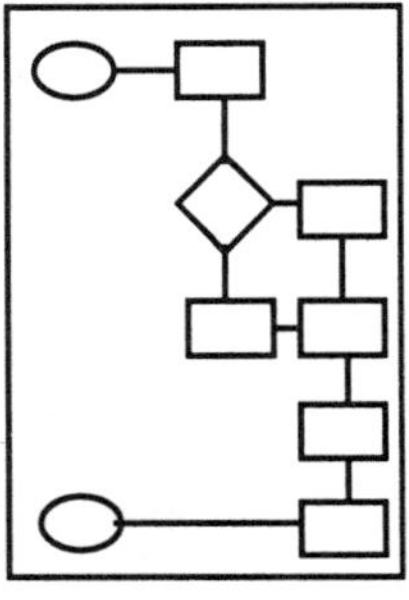

1. Brainstorm a list of potential users of the new process. Anyone who shares a similar customer, product, or service could potentially benefit from the new process.

2. Select the key groups to replicate the process.

3. Forward the process (flowchart and indicators) to the groups targeted for replication.

4. Follow up to ensure that the improvements have been implemented. Replicate any additional enhancements that have been made by these other groups.

WHERE? Where will this process be useful?	WHAT? What needs to be done to initiate?	HOW? How will the process be replicated?	WHO? Who owns the replication?	WHEN? Start	Complete
		Adopt process Adapt process to fit Incorporate existing improvements			

The QI Connect-the-Dots Book